Dimitri Plemenos and Georgios Miaoulis (Eds.)

Artificial Intelligence Techniques for Computer Graphics

# Studies in Computational Intelligence, Volume 159

**Editor-in-Chief**

Prof. Janusz Kacprzyk
Systems Research Institute
Polish Academy of Sciences
ul. Newelska 6
01-447 Warsaw
Poland
*E-mail:* kacprzyk@ibspan.waw.pl

Dimitri Plemenos
Georgios Miaoulis
(Eds.)

# Artificial Intelligence Techniques for Computer Graphics

 Springer

Professor Dimitri Plemenos
Universite de Limoges
Laboratoire XLIM
83,rue d'Isle
87000 Limoges
France
Email: plemenos@unilim.fr

Professor Georgios Miaoulis
Technological Education Institute of Athens
Department of Computer Science
Ag. Spyridonos
Egaleo, 122 10 Athens
Greece

ISBN 978-3-540-85127-1                    e-ISBN 978-3-540-85128-8

DOI 10.1007/978-3-540-85128-8

Studies in Computational Intelligence          ISSN 1860949X

Library of Congress Control Number: 2008933224

© 2008 Springer-Verlag Berlin Heidelberg

*Typeset & Cover Design:* Scientific Publishing Services Pvt. Ltd., Chennai, India.

Printed in acid-free paper

9 8 7 6 5 4 3 2 1

springer.com

# Preface

The purpose of this volume is to present current work of the Intelligent Computer Graphics community, a community growing up year after year. Indeed, if at the beginning of Computer Graphics the use of Artificial Intelligence techniques was quite unknown, more and more researchers all over the world are nowadays interested in intelligent techniques allowing substantial improvements of traditional Computer Graphics methods. The other main contribution of intelligent techniques in Computer Graphics is to allow invention of completely new methods, often based on automation of a lot of tasks assumed in the past by the user in an imprecise and (human) time consuming manner.

The history of research in Computer Graphics is very edifying. At the beginning, due to the slowness of computers in the years 1960, the unique research concern was visualisation. The purpose of Computer Graphics researchers was to find new visualisation algorithms, less and less time consuming, in order to reduce the enormous time required for visualisation. A lot of interesting algorithms were invented during these first years of research in Computer Graphics. The scenes to be displayed were very simple because the computing power of computers was very low. So, scene modelling was not necessary and scenes were designed directly by the user, who had to give co-ordinates of vertices of scene polygons.

When, at the end of years 1970, the power of computers increased and it was possible to visualise more complex scenes, it became obvious that there were not available complex scenes and also that scene modelling is a very complex task. So, research on scene modelling began and new scene modelling techniques were proposed, as well as scene modellers allowing faster design of 3D scenes. Researchers in 3D scene modelling little by little understood that complex scene modelling is a very hard process.

In both cases, 3D scene visualisation and 3D scene modelling, problems became obvious with the need of more and more complex scenes. In 3D scene visualisation, the problem of camera position is not very hard to resolve manually when the scene to display is very simple. For this reason, Computer Graphics researchers thought that it is not a Computer Graphics problem. On the other hand, this is a very difficult and time consuming problem with complex scenes. In 3D scene modelling, if current 3D modellers allowed the designer to easily design simple scenes, it became obvious, with complex scenes, that design of complex scenes was not really computer-aided because the designer had to know and to give all details of a scene before using a scene modeller. How to create intuitive scene modellers? How to automatically estimate the camera position in scene visualisation? How, in animation, to create and animate intelligent evolving agents?

To answer all these questions, the very first research work in Intelligent Computer Graphics started at the end of 1980's. In 1994, the first 3IA Conference (3IA'94) on Intelligent Computer Graphics was organised. This volume contains both invited and selected extended papers from the last 3IA Conference (3IA'2008), together with an introduction presenting the area of Intelligent Computer Graphics and various Computer Graphics areas where introduction of intelligent techniques permitted to resolve important problems.

We hope that this volume will be interesting for the reader and that it will convince him (her) to use, or to invent, intelligent techniques in Computer Graphics and, maybe, to join the Intelligent Computer Graphics community.

Dimitri Plemenos
George Miaoulis

# Contents

# 1
# Intelligent Techniques for Computer Graphics

Dimitri Plemenos and George Miaoulis

[1] University of Limoges (France)
[2] TEI of Athens (Greece)

**Abstract.** In this chapter we present an informal definition of intelligent techniques used in Computer Graphics. These techniques may use well known Artificial Intelligence methods as well as simple human intelligence in order to automatically resolve Computer Graphics problems, usually very difficult or very fastidious to resolve by the user. Some applications of using intelligent techniques are also presented.

**Keywords:** Computer Graphics, Artificial Intelligence, Declarative Modelling, Triangulation, Camera placement, Virtual World Exploration, Motion Modelling.

## 1.1  Introduction

The use of Artificial Intelligence techniques in Computer Graphics is a quite recent research area. Even if Artificial Intelligence is used since several years to resolve problems in Image Processing, Computer Graphics remained a long time a self-sufficient science.

When, at the end of the years 1980, the research team of Dimitri Plemenos at the University of Nantes (France) started to understand the potentially rich possibilities of using Artificial Intelligence techniques in Computer Graphics and to work in this promising area, people were very skeptical and it was very difficult for this team, during at least 5 years, to publish papers on this kind of themes. People didn't understand why old techniques used in Computer Graphics had to be combined with Artificial Intelligence techniques.

The first result of this initially badly understood research work was introduction of *Declarative Scene Modelling*, that is scene modelling based on high level properties. Several Artificial Intelligence techniques, like expert systems, constraint satisfaction problem (CSP), machine-learning, neural networks and genetic algorithms were progressively integrated in the declarative modelling process.

Coming from the needs to understand scenes created using Declarative Modelling, another problem, and research area, was identified: how to automatically *choose interesting points of view*, allowing a human user to well understand a scene rendered on the screen. The next step was: how to automatically *explore virtual worlds* with a camera. In this research area, new Artificial Intelligence techniques were introduced, especially heuristic search.

D. Plemenos, G. Miaoulis (Eds.): Arti. Intel. Techn. for Comp. Graph., SCI 159, pp. 1–14.
springerlink.com                           © Springer-Verlag Berlin Heidelberg 2009

Nowadays, with the development of virtual reality, computer games and other applications on the Internet, it becomes obvious that more intuitive scene modelling and automated scene exploration with a virtual camera are more and more useful areas.

In this chapter, the general principles of some important applications of Intelligent Computer Graphics will be presented, after a section where we try to explain why Intelligent Computer Graphics is an important research area in Computer Graphics and to give an informal definition of this area.

## 1.2   Why Intelligent Computer Graphics?

The main idea is to ask the computer to do the work instead of the user. Computers work much faster than humans and human time is more precious than computer time. If the computer works more in resolution of conceptually not very difficult problems, the user will have more time for more creative tasks. Moreover, in many cases, the computer owns much more information than the user.

Let us consider a scene designer who would like to design a scene representing a house of two floors with a high tree on its left side, a swimming pool in front of the house and a small green mountain behind. Of course it is possible to design a mountain (or to, manually, find an existing model of small green mountain), then a house, to put the mountain behind the house, to design a swimming tool, to put it in front of the house and, finally, to design a high tree and to put it on the left side of the house. However, this work is very time consuming and the manner used to create the scene is not very intuitive. It is not absolutely necessary to design all the keys of tthe house before to use it in a scene. In this way the designer may loose the general image of the scene to design. Moreover, at the end of the design process, the designer obtains only one instance of the scene he (she) was in mind. If the designer wishes to compare this result with other possible results, he (she) has to start again the designing process, to choose another mountain, another house, to put then to new positions and so on.

The best solution for the designer would be to be able to tell his (her) modeller: "I would like to get a scene representing a house of two floors with a high tree on its left side, a swimming tool in front of the house and a small green mountain behind". Such a modeller should recognize high level intuitive properties like "high", "small", "green", "in front", "behind", etc. and even "house", "swimming pool", etc. and be able to propose the designer one or more solutions corresponding to these properties. Artificial Intelligence techniques may be used, mainly to reduce the processing time when looking for all scenes verifying the intuitive properties used by the designer.

Another situation, where existing classical techniques are not satisfactory, is the case where we have to display a scene and we are trying to find a point of view allowing to well see the scene. This is a problem difficult to resolve interactively, as the screen is 2-dimensional and the scene 3-dimensional. Moreover, tries are very time consuming as 3D scene rendering is generally complex.

In such a case, it is again better to ask the computer work for us. If the view quality criterion for a point of view is based on the scene geometry, the computer (more precisely the program of rendering) knows better than the user the geometry of the scene and can evaluate several points of view much faster. So, the better way to well choose a good point of view is to test "all" possible points of view and to choose the better.

Of course, it is impossible to test "all" possible points of view because there is an infinity of them and, for this reason, Artificial Intelligence techniques like heuristic search may be used in order to test only "pertinent" points of view.

Many other areas of Computer Graphics are concerned by the idea to confide to the computer more and more tasks and to automatise time consuming interactive processes. Our purpose is to simplify the work of the user by allowing intuitive resolution of Computer Graphics problems. Artificial as well as human Intelligence is used in order to find practical solutions to general or specific problems.

## 1.3  Declarative Scene Modelling

*Declarative scene modelling* [1, 2, 3, 5] in computer graphics is a very powerful technique allowing to describe the scene to be designed in an intuitive manner, by only giving some expected properties of the scene and letting the modeller find solutions, if any, verifying these properties.

As the user may describe a scene in an intuitive manner, using common expressions, the described properties are often imprecise. For example, the user can tell the modeller that "the scene A must be put on the left of scene B". There exist several possibilities to put a scene on the left of another one. Another kind of imprecision is due to the fact that the designer does not know the exact property his (her) scene has to satisfy and expects some proposals from the modeller. So, the user can indicate that "the house A must be near the house B" without giving any other precision. Due to this lack of precision, declarative modelling is generally a time consuming scene modelling technique.

It is generally admitted that the declarative modelling process is made of three phases: the *description* phase, where the designer describes the scene, the *scene generation* phase, where the modeller generates one or more scenes verifying the description, and the *scene understanding* phase, where the designer, or the modeller, tries to understand a generated scene in order to decide whether the proposed solution is a satisfactory one, or not.

Several general purpose or dedicated declarative scene modeller were developed, especially in France.

**PolyFormes** is the first experimental declarative scene modeller. It was developed in Nantes (France). The goal of the PolyFormes declarative modeller is to generate all regular and semi-regular polyhedra, or a part of the whole, according to the user's request [4]. The whole modeller is an expert system on polyhedra. When the initial description is imprecise, all the possible solutions are generated. Fig. 1.1 shows an example of polyhedron generated by PolyFormes.

**MultiFormes** is a general purpose declarative scene modeller, based on a new conception and modelling technique, declarative modelling by hierarchical decomposition (DMHD) . The DMHD technique can be resumed as follows [2, 3]:

- If the current scene can be described using a small number of predefined high level properties, describe it.
- Otherwise, describe what is possible and then decompose the scene in a number of sub-scenes. Apply the DMHD technique to each sub-scene.

**Fig. 1.1.** Example of polyhedron generated by the PolyFormes declarative scene modeller

The tree of the hierarchical description of a scene, used in the scene generation phase, allows scene generation in various levels of detail and reduction of the generation's cost. To do this, the modeller uses a bounding box for each node of the tree. This bounding box is the bounding box of the sub-scene represented by the sub-tree whose the current node is the root. All bounding boxes of the children nodes of a node are physically included in the bounding box of the parent node. This property permits to detect very soon branches of the generation tree which cannot be solutions. In Fig. 1.2, the spatial relation between the bounding boxes of a scene and its sub-scenes is shown (left), as well as two scenes generated by MultiFormes (middle and right).

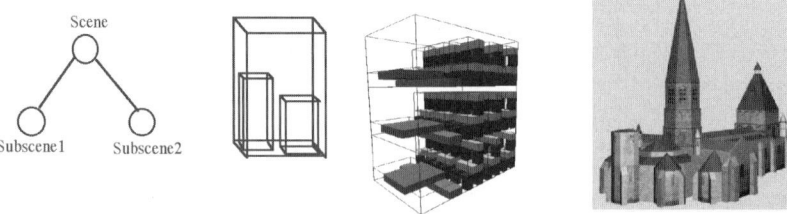

**Fig. 1.2.** From left to right: the bounding boxes of the sub-scenes of a scene are inside the bounding box of the parent scene; inside a 3-floor building; Cathedral of Le Dorat (France) designed by W. Ruchaud

**DE2MONS** [6] is a general purpose modeller whose main properties are:

- A multi modal interface,
- A generation engine limited to the placement of objects,
- A constraint solver able to process dynamic and hierarchical constraints.

The modeller uses a multi modal interface allowing descriptions by means of the voice, the keyboard (natural language), a data glove or 3D captors informing the system of the user's position. The description is translated in an internal model made of linear constraints.The generation engine of DE2MONS uses a linear constraint solver, ORANOS, able to process dynamic constraints (new constraints can be added during generation) and hierarchical constraints. Hierarchical constraints are constraints with priorities assigned by the user. Whenever there is no solution for a given description, constraints with low priority are released in order to always get a solution. The solver computes one solution for a given description.

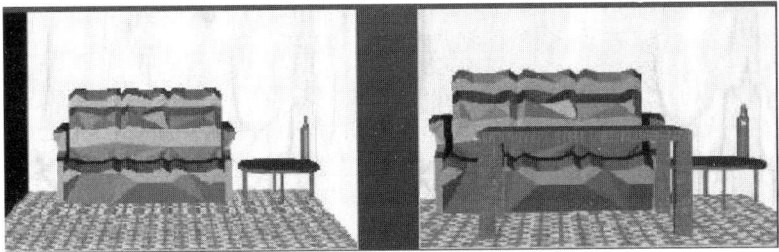

**Fig. 1.3.** Furniture pieces placement with DE2MONS

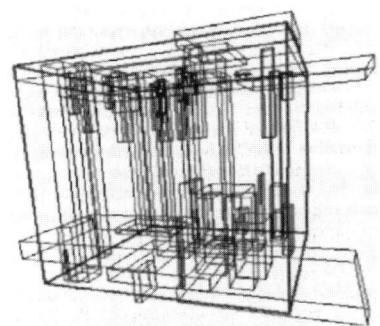 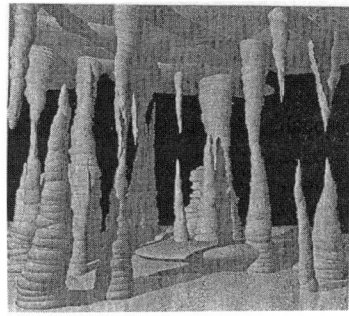

**Fig. 1.4.** Boxes arrangement and form growth with VoluFormes

Images in Fig. 1.3 show furniture pieces placements generated by the generation engine of DE2MONS.

**VoluFormes** [7] is a *dedicated* declarative modeller allowing the user to quickly define boxes in the space whose purpose is to check the growth of forms. It is made of two modules:

- *Voluboites*, which allows to define boxes where the spatial control is performed.
- *Voluscenes*, which allows to use growth mechanisms applied to elementary germs and to create forms, taking into account the spatial control boxes.

On the left of Fig. 1.4, one can see a boxes arrangement obtained by Voluboites, while on the right a result of application of the form growth mechanism is presented.

## 1.4  Visual Scene Understanding

One of the main goals of 3D scene rendering is to understand a scene from its rendered image. However, the rendered image of the scene depends on the choice of the point of view. If the point of view is a bad one, the scene may be very difficult to understand from its image. Interactively choosing a point of view for a scene is a very difficult and human time consuming process because on the one hand the user has not a good knowledge of the scene and, on the other hand, it is not easy for the user to select a point of view from a 3D scene from a 2D screen.

As the computer has a more complete knowledge of the scene, it is much better to ask the computer to find an interesting point of view. To do this, it is necessary to define a view quality criterion, in order to give the computer a way to evaluate points of view and to choose the better one.

As there is an infinity of potential points of view for a given scene, we also need a way to evaluate only a finite number of points of view and, even, to avoid evaluation of useless ones.

The first general method of automated computation of a good viewpoint for a polygonal scene was proposed in 1991 [2] and improved in 1996 [8]. In this method, the main viewpoint quality criterion is the number of visible polygons of the scene from

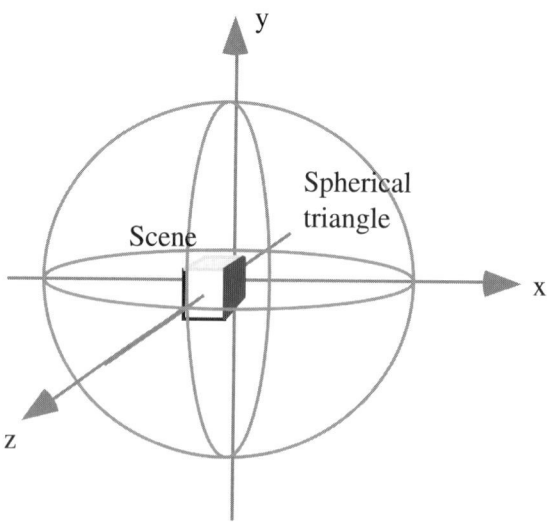

**Fig. 1.5.** The scene is at the centre of a surrounding sphere

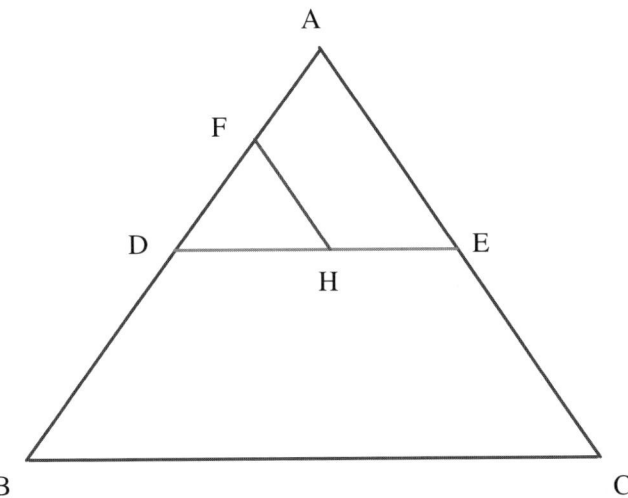

**Fig. 1.6.** From an initial spherical triangle ABC, a new spherical triangle ADE is computed and so on. The vertices of spherical triangles represent points of view.

the given point of view, together with the total visible area of the scene projected on the screen. All potential viewpoints are supposed to be on the surface of a sphere surrounding the scene (Fig. 1.5). Heuristic search is used to choose viewpoints only in potentially interesting regions, obtained by subdivision of spherical triangles (Fig. 1.6).

A lot of other methods were proposed during this last decade [9,12, 18, 19]. Some recent methods try to take into account not only the geometry but also the lighting of the scene in order to evaluate the quality of a point of view [10, 14].

## 1.5  Virtual World Exploration

When a virtual world is complex, it is often difficult and even impossible for the user to understand it from a single image, that is from a single point of view. Show a set of images of the virtual world, instead of only one, does not meaningfully improve understanding because it may be confusing for the user.

The best solution for optimized understanding of complex virtual worlds is a continuous exploration with a camera. The camera may start exploration from the best point of view and then move around (or inside) the virtual world, choosing good points of view for its movement and avoiding brusque changes of direction. Fig. 1.7 shows an example of incremental exploration of an office by a virtual camera.

Several papers were published these last years on virtual world exploration [9, 13, 18].

Virtual world exploration may be incremental online exploration, where the camera has to use only local knowledge in order to get real time exploration, or offline exploration, where the camera may use global knowledge, in order to get more precise exploration when, later, the user would wish to explore the virtual world.

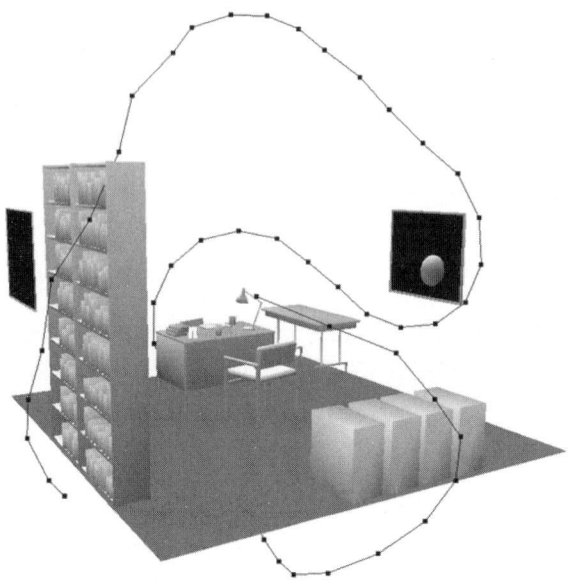

**Fig. 1.7.** Incremental exploration of an office

## 1.6  Behavioral Animation

In virtual reality applications, autonomous virtual agents are entities which own their proper behaviour and to which individual tasks are assigned. They have to move and to take decisions accordingly to their tasks and depending on the current situation. In other words, they must be able to make intelligent exploration of their environment.

Virtual agents must be able to get information on the environment and then [16]:

1.    Transform this information in knowledge.
2.    Use this knowledge to take decisions
3.    Make the decided actions.

Depending on the mechanisms used to store knowledge and to take decisions from this knowledge various Artificial Intelligence techniques are used such as expert systems, genetic algorithms, neural networks etc.

## 1.7  Getting Optimal Triangulated Models

In Computer Graphics triangulated models are abundantly used because of their simplicity and their hardware implementation in graphical cards. However, these models are often difficult to obtain. If for 2D data (sets of  2D points) there exist planar triangulation algorithms, in the 3D case there not exist any algorithm, except for special cases.

Even with classical planar triangulation algorithms the obtained triangulated models are not optimal. In order to increase optimality of obtained triangulated models, some Artificial Intelligence techniques are used, especially genetic algorithms and simulated annealing, or a combination of these techniques. Current results are not fully satisfactory, mainly due to the time and memory needs of these techniques, especially of genetic algorithms [15].

## 1.8   Intelligent Image-Based Modelling and Rendering

In image-based modelling, it is important to compute a minimum set of points of view in order to use them in obtaining the set of images which will replace the scene display. A method to do this, based on viewpoint complexity, is the following [ 19]:

1. Compute a sufficient number of points of view, according to the viewpoint complexity of the scene from a region of the sphere of points of view. A region is represented by a spherical triangle. The viewpoint complexity of a scene from a region may be defined as the average value of viewpoint complexities from the three vertices of the spherical triangle.
   - If the scene is viewpoint complex from the region, subdivide the region in 4 sub-regions.
   - Otherwise, add the centre of the region to the current list of point of view.
2. Using an evaluation function for a set of points of view, try to eliminate elements of the list of points of view in several steps, by considering each point of view as a candidate to be eliminated. A point of view is eliminated if its contribution is empty (or quasi-empty) to the result of evaluation of the remaining point of view.
3. If, at the end of a step, no one point of view is eliminated, the process is finished and the current list of point of view contains a minimum set of points of view.

## 1.9   Inverse Lighting

Inverse lighting techniques are used to answer the general question: For a given scene, determine the number n of light sources S1, S2, ..., Sn, their corresponding areas A1, A2, ..., An and positions P1, P2, ..., Pn, in order to get a wished lighting ambience at a chosen part of the scene.

In order to describe a wished ambience, declarative modeling high level properties may be used, together with fuzzy logic. Constraint satisfaction techniques are used to find solutions [17].

In Fig. 1.8 and Fig. 1.9 one can see examples of ambience modelling in a foggy landscape. These examples present two solutions of ambiance for the wished high level property "Visibility is of 70 meters".

**Fig. 1.8.** A solution for the property "Visibility is of 70 meters"

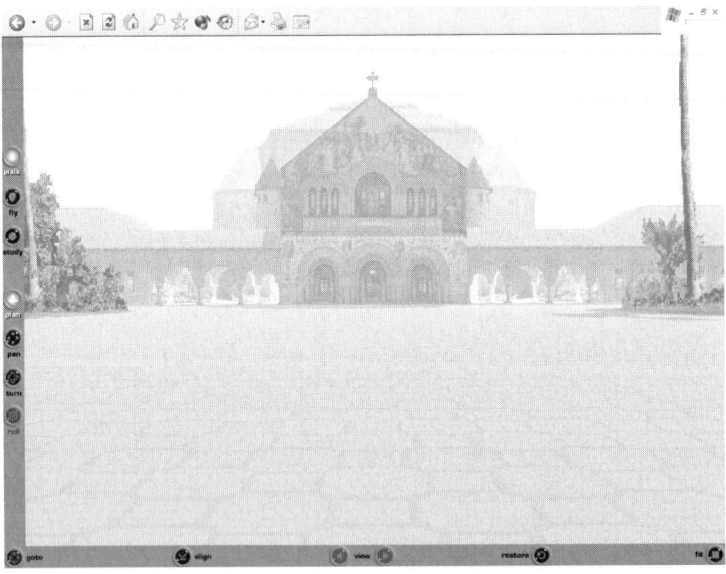

**Fig. 1.9.** Another solution for the property "Visibility is of 70 meters"

## 1.10  Alternate Visualization Techniques

In some cases it is necessary to use simple human intelligence in order to resolve particular problems in Computer Graphics. Most of these cases concern understanding of

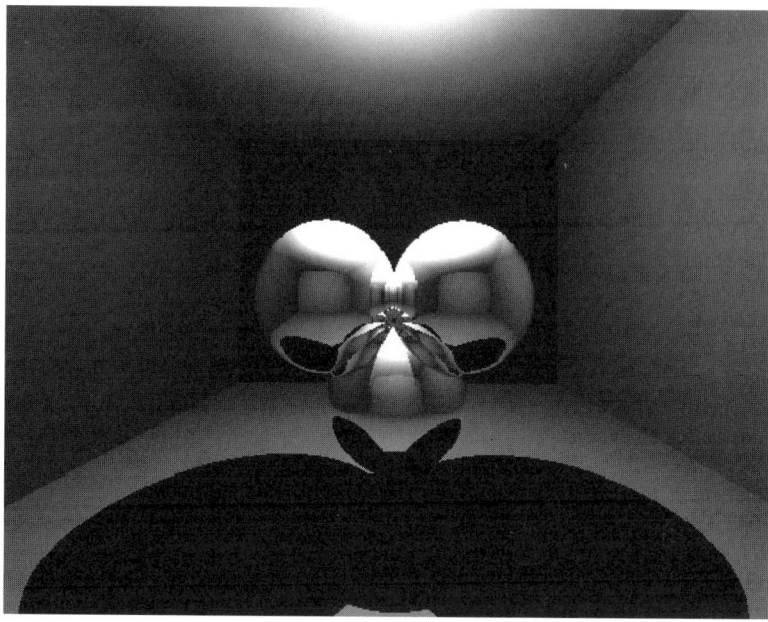

**Fig. 1.10.** Scene with many light effects, difficult to understand

**Fig. 1.11.** The real scene is delimited by drawing the contours of its elements

special kinds of scenes. These problems may be resolved by using alternate visualization techniques.

A first example of special kind of scene is the one where the scene contains mirrors and/or several light sources. Many of these scenes are difficult to understand from a single image because reflections and shadows create illusions which are very often impossible to distinguish from the real scene. An example of such a scene is shown in Fig. 1.10.

Fig. 1.11 shows an example of alternate visualization technique, where the contours of the real part of the scene are extracted and drawn [11].

Another example of scene requiring special visualization is the one of complex enough scenes which have to be understood from a single image. Sometimes, in newspapers, one can see a picture of plane, seen from outside, where an artificial hole allows to see some parts of the inside of the plane too, for example some seats. How to get automatically this such images? The best solution, if we know which parts inside the scene we have to make visible, is to apply a preprocessing step, where polygons, or parts of polygons, obstructing the view of objects which must be visible, are removed from the scene before running a hidden surface algorithm.

## 1.11   The 3IA International Conference

In 1994, Dimitri PLEMENOS organised in Limoges the first 3IA Conference on Computer Graphics and Artificial Intelligence. This Conference became international in 1996. During the first years of the 3IA Conference the dominant theme was declarative modelling. Then, year after year, more and more people used Artificial Intelligence techniques in various computer graphics areas and many other interesting applications are proposed every year.

Nowadays the 3IA International Conference became the annual meeting of researchers all over the world, using intelligent techniques to improve Computer Graphics techniques. This year, accepted papers in the 3IA'2008 International Conference were submitted from 20 different countries. Selected papers for this volume cover several different areas: Virtual reality, motion modelling, object retrieval, intelligent modelling and rendering, declarative modelling, genetic algorithms and data visualization.

For more information on the 3IA International Conference the lector is invited to visit the current web site of the Conference: http://3ia.teiath.gr.

## 1.12   Conclusion

In this chapter we tried to explain the concept of Intelligent Computer Graphics. Our goal was to show how Artificial Intelligence, and even simple human intelligence, may improve current Computer Graphics techniques and greatly simplify the user's work.

Even if it is not exhaustive, this presentation of numerous Computer Graphics areas where the use of intelligent techniques is today a reality demonstrates the vitality of this tendency. Many other applications of intelligent techniques to resolve

Computer Graphics problems are currently investigated and interesting results are presented almost every day.

In many cases, the main contribution of the use of intelligent techniques in Computer Graphics is automation of a lot of processes, which in the past had to be performed manually by the user. In other cases, where intelligent techniques are used in non interactive processes, their use allow to invent new original methods to resolve Computer Graphics problems. Both kinds of contributions are very important.

# References

1. Lucas, M., Martin, D., Martin, P., Plemenos, D.: The ExploFormes project: some steps towards declarative modelling of forms. BIGRE, n° 67, pp. 35–49 (1990)
2. Plemenos, D.: A contribution to study and development of scene modelling, generation and display techniques – The MultiFormes project. Professorial dissertation, Nantes (France) (November 1991)
3. Plemenos, D.: Declarative modelling by hierarchical decomposition. The actual state of the MultiFormes project. In: International Conference GraphiCon 1995, Saint Petersburg (Russia), July 3-7 (1995)
4. Martin, D., Martin, P.: PolyFormes software for the declarative modelling of polyhedra. The Visual Computer, 55–76 (1999)
5. Bonnefoi, P.F., Plemenos, D., Ruchaud, W.: Declarative modelling in computer graphics: current results and future issues. In: ICCS 2004 international conference (CGGM 2004). LNCS, Krakow (Poland), June 6-9, pp. IV80–IV89. Springer, Heidelberg (2004)
6. Kwaiter, G.: Declarative scene modelling: study and implementation of constraint solvers. Ph.D thesis, Toulouse (France) (December 1998)
7. Chauvat, D.: The VoluFormes project: An example of declarative modelling with spatial control. Ph.D thesis, Nantes (France) (December 1994)
8. Plemenos, D., Benayada, M.: Intelligent Display in Scene Modeling. New Techniques to Automatically Compute Good Views. In: International Conference GraphiCon 1996, St Petersbourg (Russia), July 1-5 (1996)
9. Sokolov, D., Plemenos, D.: Virtual world exploration by using topological and semantic knowledge. The Visual Computer 24(3) (March 2008)
10. Lam, C., Plemenos, D.: Intelligent scene understanding using geometry and lighting. In: 10th International Conference 3IA 2007, Athens (Greece), May 30-31, pp. 97–108 (2007)
11. Dandachy, N., Plemenos, D., El Hassan, B.: Scene understanding by apparent contour extraction. In: 10th International Conference 3IA 2007, Athens (Greece), May 30-31, pp. 85–96 (2007)
12. Vázquez, P.P., Feixas, M., Sbert, M., Heidrich, W.: Viewpoint Selection Using Viewpoint Entropy. In: Vision, Modeling, and Visualization 2001 (Stuttgart, Germany), pp. 273–280 (2001)
13. Vázquez, P.P., Sbert, M.: Automatic indoor scene exploration. In: International Conference on Artificial Intelligence and Computer Graphics, 3IA, Limoges (May 2003)
14. Vázquez, P.P., Sbert, M.: Perception-based illumination information measurement and light source placement. In: Kumar, V., Gavrilova, M.L., Tan, C.J.K., L'Ecuyer, P. (eds.) ICCSA 2003. LNCS, vol. 2669. Springer, Heidelberg (2003)
15. Kolingerova, I.: Probabilistic methods for triangulated models. In: 8th International Conference 3IA 2005, Limoges (France), May 11-12, pp. 93–106 (2005)

16. Iglesias, A., Luengo, F.: Applying artificial intelligence techniques for behavioral animation of virtual agents. In: 7th International Conference 3IA 2004, Limoges (France), May 12-13, pp. 59–71 (2004)
17. Jolivet, V., Plemenos, D., Poulingeas, P.: Declarative specification of ambiance in VRML landscapes. In: ICCS 2004 international conference CGGM 2004. LNCS, Krakow (Poland), June 6-9, pp. IV115–IV122. Springer, Heidelberg (2004)
18. Sokolov, D., Plemenos, D., Tamine, K.: Methods and data structures for virtual world exploration. The Visual Computer 22(7), 506–516 (2006)
19. Plemenos, D., Sbert, M., Feixas, M.: On viewpoint complexity of 3D scenes. STAR Report. In: International Conference GraphiCon 2004, Moscow (Russia), September 6-10, pp. 24–31 (2004)

# 2

# Synthesizing Human Motion from Intuitive Constraints

Alla Safonova[1] and Jessica K. Hodgins[2]

[1] University Of Pennsylvania, USA
   alla@cis.upenn.edu
[2] Carnegie Mellon University, USA
   jkh@cs.cmu.edu

**Summary.** Many compelling applications would become feasible if novice users had the ability to synthesize high quality human motion based only on a simple sketch and a few easily specified constraints. Motion graphs and their variations have proven to be a powerful tool for synthesizing human motion when only a rough sketch is given. Motion graphs are simple to implement, and the synthesis can be fully automatic. When unrolled into the environment, motion graphs, however, grow drastically in size. The major challenge is then searching these large graphs for motions that satisfy user constraints. A number of sub-optimal algorithms that do not provide guarantees on the optimality of the solution have been proposed. In this paper, we argue that in many situations to get natural results an optimal or nearly-optimal search is required. We show how to use the well-known A* search to find solutions that are optimal or of bounded sub-optimality. We achieve this goal for large motion graphs by performing a lossless compression of the motion graph and implementing a heuristic function that significantly accelerates the search for the domain of human motion. We demonstrate the power of this approach by synthesizing optimal or near optimal motions that include a variety of behaviors in a single motion. These experiments show that motions become more natural as the optimality improves.

## 2.1 Introduction

The ability to construct animations of human characters easily and without significant training would enable many novel and compelling applications. Children could animate stories, novice users could author effective training scenarios, and game players could create a rich set of character motions. With these applications in mind, we have focused on techniques that require users to provide only a small amount of information about a desired motion. The user provides an approximate sketch of the path of the character on the ground plane and a set of constraints (Figure 2.1). Optimization is a common technique for finding a motion when only a rough sketch is provided. A number of continuous optimization techniques have been proposed for solving this problem (for example, [5, 19, 20, 22]).

D. Plemenos, G. Miaoulis (Eds.): Arti. Intel. Techn. for Comp. Graph., SCI 159, pp. 15–39.
springerlink.com                                    © Springer-Verlag Berlin Heidelberg 2009

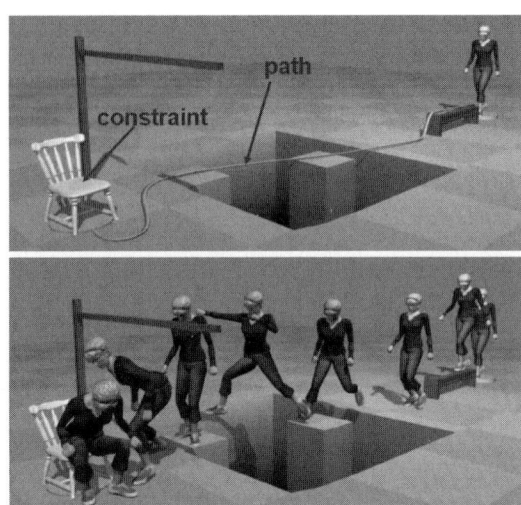

**Fig. 2.1.** (Top) Rough sketch of the desired path and a user constraint requiring the character to sit on the chair at the end of the path. (Bottom) Synthesized motion.

In this paper, we concentrate on discrete optimization techniques that search a graph constructed from existing motion capture data for the solution.[1]

Motion graphs have proven to be a a powerful technique to solve for a desired motion based only on a rough sketch [1, 2, 7, 10, 12]. Because the solution is constrained to a sequence of motion segments from the motion capture database, motion graphs are restrictive in the set of motions they can represent. For example, it would be impossible to synthesize a motion for picking up a cup from a table that is 1.0 meter high if the database contains only motions for picking up a cup from tables that are 0.5 and 1.5 meters high. To relax this restriction, in [18] we have introduced interpolated motion graphs (IMG). The motion is represented as an interpolation of two time-scaled paths through a motion graph. The strength of this representation is that it allows the adaptation of existing motions through interpolation while also retaining the natural transitions present in a motion graph. Although larger than a motion graph, this representation creates a search space that is far smaller than the full space created by the 50 degrees of freedom of a human character because it contains only natural poses and velocities from the original motions and the interpolation of segments with matching contact patterns.

Discrete search techniques can be used to search a motion graph or an interpolated motion graph for a motion that satisfies user-specified constraints. During the search, the graph is unrolled into the environment (by augmenting

---

[1] This work is based on an earlier work: Construction and optimal search of interpolated motion graphs, in ACM SIGGRAPH 2007 Papers ©ACM, 2007. http://doi.org/10.1145/1275808.1276510

each state with the global position and orientation of the root). This step is required to search for motions that satisfy user-specified global position constraints and avoid obstacles. Unrolling causes the size of the graph to grow drastically in size and makes search challenging.

Most motion graph implementations have used sub-optimal techniques (that do not provide guarantees on the optimality of the solution) because it was thought to be infeasible to perform a global search of a motion graph of sufficient size to produce natural motion. In this paper, we argue that in many situations to get natural results an optimal or nearly-optimal search is required.

We show how to use the well-known $A^*$ search (we use an anytime version of $A^*$ by Likhachev et. al [13]) to find solutions that are optimal or of bounded sub-optimality in a motion graph containing a variety of behaviors. $A^*$ is a breadth-first algorithm that gains its efficiency by using a lower bound on the cost to the goal (a heuristic) to avoid computing nodes that are irrelevant to the optimal plan.

We were able to use $A^*$ on a motion graph because we made two improvements to the standard motion graph implementation. The first technique compresses the motion graph into a *practically equivalent* but much smaller graph by removing states and transitions that would not be part of an optimal solution or are redundant. The compression reduces the size of the graph by a factor of 20 to 50.

The second technique computes an informative heuristic function that guides the search toward states that are more likely to appear in an optimal solution. We created the heuristic by splitting the full search problem into two much simpler problems: one that is only concerned with the character's location and orientation in the environment and the other based purely on the motion graph with no environmental constraints. Both of these smaller planning problems can be solved efficiently. The solutions to these problems provide lower bounds on the costs to the goal that can be combined to obtain an informative heuristic function for the search.

The combination of these two techniques makes it possible to find optimal or close-to-optimal solutions in standard motion graphs at close to interactive rates (10 seconds of motion required a few seconds of computation time for our examples). It also makes it possible to search interpolated motion graphs for an up to 15 second motion with a few minutes of computation.

In our experiments the user specified a desired 2D path that required the character to perform various behaviors such as jumping, walking, walking along a beam, ducking, picking, sitting. The solution minimizes the sum of squared accelerations (an approximation to energy) and therefore our implementation avoids the dithering back and forth motions often seen in solutions computed with a local or sub-optimal search [10]. The optimality of $A^*$ allows us to find motions which are natural in that they use a running jump for longer jumps and a standing broad jump for shorter jumps as a human likely would.

## 2.2  Background

Continuous optimization, introduced to the graphics community by Witkin and Kass [22], is a common technique for finding a motion when only a rough sketch is provided. These techniques rely on physical laws to constrain the search and produce natural-looking motion. Continuous optimization has been shown to work well when a good initial guess is provided and for synthesizing relatively short, single behavior motions, such as jumps and runs (see for example [5, 19, 20]). In contrast to continuous optimization, the discrete optimization approach explored in this paper can handle longer motions that consist of multiple behaviors and does not require an initial guess.

Motion graphs and related approaches can be categorized into *on-line* approaches where the motion is generated in response to user input (from a joystick, for example) [10] and *off-line* approaches where the full motion specification is known in advance [1, 7]. On-line approaches can perform only local search because new input is continuously arriving. Off-line approaches, on the other hand, can find a high quality solution that minimizes an objective function such as energy. Our work falls into the category of off-line techniques.

A number algorithms have been developed to search a motion graph in an off-line fashion. Kovar and his colleagues [7] employed a branch and bound algorithm to get an avatar to follow a sketched path. Arikan and Forsyth [1] created a hierarchy of graphs and employed a randomized search algorithm for the synthesis of a new motion subject to user-specified constraints. Pullen and Bregler [15] segmented motion data into small pieces and rearranged them to match user-specified keyframes. In 2003, Arikan and his colleagues [2] presented a new search approach based on dynamic programming that supports user-specified annotations of the motion. The search space for their algorithm is much smaller than for ours because they do not include position information with each state but only the character's pose and a time. This simplification makes it difficult to satisfy position constraints. Choi and his colleagues [4] presented a scheme for planning natural-looking locomotion of a biped figure based on a combination of probabilistic path planning and hierarchical displacement mapping. Sung and his colleagues [21] used probabilistic roadmaps and displacement mapping to synthesize motion for crowds.

To guarantee fast performance none of these approaches find optimal solutions but instead use sub-optimal search techniques. To find an optimal solution efficiently, Lau and Kuffner [9] manually created a behavior-based motion graph with a very small number of nodes. In later work, they precomputed search trees from the graph and used them for faster but not globally optimal search [8]. Lee and Lee [11] precomputed policies that indicate how the avatar should move for each possible control input and avatar state. Their approach allows interactive control with minimal run-time cost for a restricted set of control inputs.

Unlike all previous motion graph approaches with the exception of Lau and Kuffner [9], we find a globally optimal or a close-to-optimal solution with an upper bound on the sub-optimality. In Section 2.6, we show a number of comparisons to demonstrate that globally optimal solutions avoid the inefficient patterns of motion that are often seen with local or sub-optimal search techniques.

## 2.3 Overview

We assume that we have a database of motions sampled as an ordered sequence of poses. We use a right-handed coordinate system $XYZ$ with the $X$ and $Z$ axes spanning the ground plane and the $Y$ axis pointing up. Each pose is represented by (1) $Q$, the joint angles relative to the inboard link and the orientation of the root around the $X$ and $Z$ axes, (2) $P_y$, the position of the root along the vertical axis, (3) $\Delta P_x$ and $\Delta P_z$, the relative position of the root on the ground plane (computed with respect to the previous pose in this motion sequence) and (4) $\Delta Q_{yaw}$, the relative rotation of the root around the vertical axis (computed similarly).

The goal is to find motion that satisfies user sketch and constraints and at the same time is natural and physically correct. Optimization is a common technique for finding a motion when only a rough sketch is provided. The user specifies a set of constraints (such as pose) and an objective function. The optimization problem is then to minimize the objective function while satisfying user-specified and physics constraints (which preserve the physical validity of the motion). In this paper we concentrate on discrete optimization techniques.

To setup optimization problem we need to define unknown variables that optimizer will search for, constraints that optimizer will need to satisfy and the objective function that optimizer will minimize to find a natural solution. In Section 2.4 we describe each of them.

To solve the optimization problem we use a database of motions to construct a graph that will be searched for an optimal solution. A number of different graphs have been proposed in the literature, including behavior-based graphs, motion graphs and graphs that combine motion graphs with interpolation. We describe graph construction process in Section 2.5.

A number of search algorithms can be used to search constructed graph for a solution. Global search methods compute the whole solution in one search. Local search methods, on the other hand, only perform short horizon searches that repeatedly find few steps of the solution. Global methods find better quality solutions at the expense of longer computation times. Global search methods can be divided into methods that find a globally optimal solution and ones that find a locally optimal solution with no guarantee on sub-optimality. We discuss the importance of finding an optimal solution in Section 2.6.

In Section 2.7 we show how to use the well-known $A^*$ search (we use an anytime version of $A^*$ [13]) to find solutions that are optimal or of bounded sub-optimality. We achieve this goal for large graphs by performing a lossless compression of the graph and implementing a heuristic function that significantly accelerates the search for the domain of human motion. We present our results in Section 2.8 and conclude with the discussion in Section 2.9.

## 2.4   Discrete Optimization Problem

**Unknowns:** Given user sketch and constraints (such as one shown in Figure 2.1), the goal is to find the motion of the character that stays close to user sketch and satisfies all constraints. Therefore, unknowns of the optimization problem are poses of the character over time.

**Objective function:** Variety of objective functions have been proposed in the literature. For example, in their work [1], Arikan and Forsyth, use an objective function that is a summation of violations of soft constraints and smoothness of the motion. Soft constraints include: position and orientation constraints, number of frames in the resulting motion, joint constraints, style constraints and so on. Kovar and his colleges in [7], find a motion of the character that follows a sketch of the path provided by the user. The objective function they use is the difference between a user given path $P$ and the actual path traveled $P$ traversed by the character. In [2], Arikan and his colleges find a motion that satisfies given user annotations. Their objective function is a summation of $D$ abd $C$ functions, where $D$ measures the difference in annotations for each frame of the motion and $C$ measures the smoothness ("distance" between features of 2 consecutive frames).

In our work, the objective function is a weighted average of two terms: the sum of the squared torques computed via inverse dynamics and the sum of the costs of transitions associated with the traversed edges in the motion graph. The first term is an approximation of the energy needed to perform the motion. This term picks paths through the motion graph that will result in efficient motion patterns. The second term is a measure of the smoothness of the motion.

**Constraints:** Constraints allow user to express goals that need to be satisfied exactly or within small tolerance. When constraints do not need to be satisfied exactly but as close as possible, then they are treated as soft constraints and added as a term to the optimization function (as was done in [2]).

Hard constraints are most often used to specify start and goal locations of the character as was for example done in [1, 2]. Obstacle avoidance constraints are also treated as hard constraints.

In our work, all user-specified constraints are treated as constraints for the optimization problem rather than including them as part of the objective function. This decision makes the objective function independent of the particular constraints chosen by the user at runtime and allows us to compress the motion graph as a preprocessing step (Section 2.7.3). User can, for example, specify a set of constraints that constraint a particular point on the character body (such as sitting on a chair or picking an object). A user also either provides a rough sketch of the $2D$ path on the ground plane that the character should follow or the $2D$ path is computed automatically from the start and end points (Figure 2.2). The root of the character is constrained to stay inside a $2D$ corridor around the path ($0.5 - 1.0$ m wide in most of the examples reported here). If the user sketch passes across obstacles (such as a river) the system also automatically adds environmental constraints (which are used for the computation of heuristic function

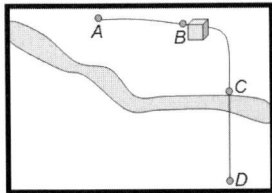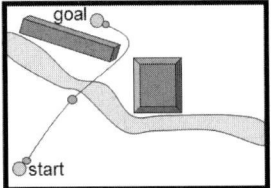

**Fig. 2.2.** Two example problem specifications. (Left) The user provided the sketch of the path of the character and specified 3 constraints: start at $A$, pick an object from a table at $B$, and arrive at $D$. An environmental constraint for jumping over the river is added automatically by the system. (Right) The user specified only the start and goal positions. The system automatically creates a sketch of the $2D$ path while avoiding obstacles and adds an environmental constraint for jumping over the river.

that guides the search). Finally, obstacle avoidance constraints are automatically included. Figure 2.2 gives an example of user sketch. User constraints should coincide with contact changes in the motion and will be met within a small tolerance.

## 2.5   Graph Construction

To solve the optimization problem we use a database of motions to construct a graph that will be searched for an optimal solution. A number of different graphs have been proposed in the literature, including behavior-based graphs, motion graphs and graphs that combine motion graphs with interpolation. In this work we show how to search standard motion graphs and interpolated motion graphs for near-optimal solutions. We give an overview of the graph construction process for these graphs next.

### 2.5.1   Motion Graphs

Motion graphs are unstructured graphs which are created completely automatically. A motion graph, $MG$, is constructed by finding "similar" poses in different motions and creating transitions between these poses (Figure 2.3). A motion can be generated by simply traversing a path in the graph. Two states are considered similar if the error between them is within some threshold. Different error functions have been shown to work (see for example [1, 7, 10]).

### 2.5.2   Interpolated Motion Graphs

Motion graphs capture natural transitions between available motion and as a result allow for creation of long, multi-behavior motions. Because the solution is constrained to a sequence of motion segments from the motion capture database motion graphs can not synthesize variations. For example, it would be impossible to synthesize a motion for picking up a cup from a table that is 1.0 meter high

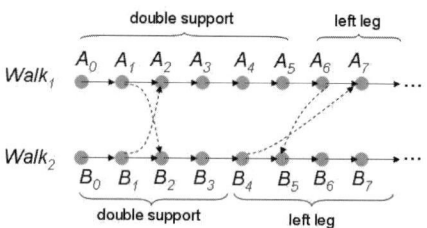

**Fig. 2.3.** A simple motion graph for two walking motions. States $A_1$ and $B_1$ are similar and therefore two transitions are added to the motion graph: $A_1 \rightarrow B_2$ and $B_1 \rightarrow A_2$.

if the database contains only motions for picking up a cup from tables that are 0.5 and 1.5 meters high.

To relax this restriction, we have introduced interpolated motion graph, $IMG$ ( [18]). The key insight behind interpolated motion graphs is that the motion is represented as an interpolation of two time-scaled paths through a motion graph. The strength of this representation is that it allows the adaptation of existing motions through interpolation while also retaining the natural transitions present in a motion graph. We allow interpolation only of segments with matching contact patterns and therefore the resulting motion is often close to physically correct [17]. Although larger than a motion graph, this representation creates a search space that is far smaller than the full space created by the 50 degrees of freedom of a human character because it contains only natural poses and velocities from the original motions and the interpolation of segments with matching contact patterns.

We represent the motion, $M'(t)$, that we are trying to synthesize as an interpolation of two time-scaled paths through a motion graph:

$$M'(t) = w(t)M_1(t) + (1 - w(t))M_2(t). \tag{2.1}$$

where $M_1(t)$ and $M_2(t)$ are the paths and $w(t)$ is an interpolation weight. The two paths independently transition between poses in the database (Figure 2.4). We allow paths to be scaled in time to synchronize the motions for interpolation. The weight, $w(t)$ can also change with time. Equation 2.1 is very similar to the standard equation for motion interpolation, where $M_1(t)$ and $M_2(t)$ are two short motion segments of similar structure (two jumps, for example). In our representation $M_1(t)$ and $M_2(t)$ are two long paths through the motion graph which we find using discrete optimization.

We construct a graph that supports interpolation of paths through the original motion graph. We first construct a standard motion graph, $MG$ as described in the previous section. We construct graph $MG$ as a preprocessing step. We then generalize $MG$ to create a motion graph that supports interpolation. We call this graph $IMG$ (interpolated motion graph). Graph $IMG$ can also be constructed as a preprocessing step because it does not require significant space (for our examples $IMG$ would require less than $5MB$).

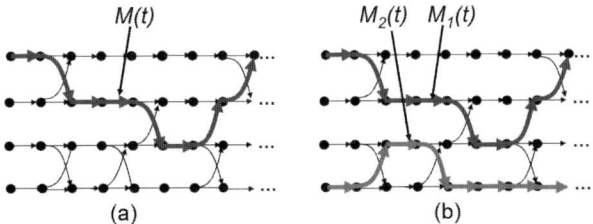

**Fig. 2.4.** For this example, the database consists of four motions: two walks and two jumps. (a) Standard motion graphs find one path through the graph; (b) Interpolated motion graphs find two paths through the graph. The resulting motion is an interpolation of these two paths, $M_1(t)$ (red) and $M_2(t)$ (pink). $M_1(t)$ and $M_2(t)$ can transition independently between motions in the database.

Each state in graph $IMG$ is defined as $S = (I_1, I_2, w)$, where $I_1$ and $I_2$ are the indices of the two poses to be interpolated and $w$ is the interpolation weight. Constructing graph $IMG$ is like taking the "product" of two identical motion graphs. Thus, the maximum number of states in graph $IMG$ is $N^2W$, where $N$ is the number of poses in the motion capture database and $W$ is the number of possible weight values. In practice, however, the number of states is much smaller because we interpolate only poses with matching contact states (left foot on the ground, for example). Given state $A$ defined by $(I_1^A, I_2^A, w_1^A)$, we need to compute the set of successor states—the states that can be reached from state $A$ via a transition in the graph $IMG$. State $B$ is a successor of state $A$ if and only if $I_1^B$ is a successor of $I_1^A$, and $I_2^B$ is a successor of $I_2^A$ in the motion graph $MG$.

### 2.5.3 Unrolling into Environment

During the search, the graph is unrolled into the environment (by augmenting each state with the global position and orientation of the root). We call unrolled graph - $SG$ (search graph). This step is required to search for motions that satisfy user-specified global position constraints and avoid obstacles. The process of unrolling is the same for both, standard motion graphs and interpolated motion graphs. In this section, we use motion graphs to describe unrolling.

This graph is built as the search works on it, thereby limiting the allocated memory to only the states that are actually computed.Each state in the graph $SG$ is defined by a pose in the motion graph $MG$ and a global position and orientation of the root of the character in the environment. The global position is required in order to satisfy position constraints (such as getting to the goal). Depending on the problem, other variables could be added to each state (elapsed time, for example). Unfortunately, the addition of the position information increases the number of possible states significantly. For example, suppose the position of a character is defined by a $2D$ position in the plane ($x$ and $z$) and an orientation about vertical axis ($\theta$). Suppose also that each of these variables can take 1000 distinct values. If the motion graph $MG$ has $10^4$ states then the

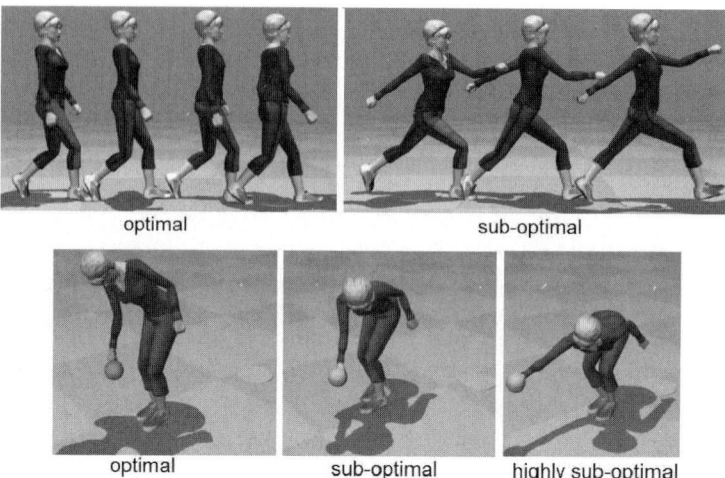

optimal                     sub-optimal

optimal          sub-optimal          highly sub-optimal

**Fig. 2.5.** Optimal and sub-optimal solutions for walking a given distance (top) and for picking up an object (bottom)

search graph $SG$ will have $10^4 * 10^9$ states. This exponential expansion makes search in this graph difficult if not impossible.

During the search, in addition to augmenting each state with root position and orientation, we also augment each state with a constraint counter. The counter is used to ensure that all constraints are satisfied. If a state satisfies the next constraint during a search, its counter is set to that of its predecessor plus one. Because constraints are positioned along the user sketch, the counter also allows states to be pruned from the search if they pass a constraint without satisfying it.

## 2.6 Importance of Optimality

The search can be global but with no guarantees on optimality of the solution. In many cases, however, it is important to find a solution as close to an optimal as possible. Globally near-optimal solutions avoid the dithering and inefficient patterns of motion that greedy or locally optimal solutions often have [10].

Figure 2.5 shows the results for two motions: a walk to a place where the character needs to pick up an object and a walk from the start to the goal. As the optimality of the solution increases, the character finds more efficient motion patterns. Figure 2.6 shows jumping example. Optimal solution uses a running jump rather than a standing broad jump as a human likely would.

Tables 2.1 shows how the cost of the solution changes as its optimality increases during our search. Top table is for "walk from start to goal" example.The first solution is very suboptimal—the character makes two really large steps to reach the goal position (Figure 2.5). The second solution is better—the character makes smaller steps but the walk is a bit unnatural because the steps are of different length. The final solution is optimal and looks natural (Figure 2.5).

**Fig. 2.6.** Optimal (right) and sub-optimal (left) solutions for jumping over river

**Table 2.1.** Importance Of Optimality

| Search Time (secs) | Solution Cost | Optimality Bound ($\epsilon$) |
|---|---|---|
| 0.67 | 2,700,000 | 10.0 |
| 3.42 | 1,800,000 | 2.0 |
| 50.00 | 1,600,000 | 1.0 |

| Search Time (secs) | Solution Cost | Optimality Bound ($\epsilon$) |
|---|---|---|
| 0.016 | 10,700,000 | 10.0 |
| 0.032 | 8,200,000 | 2.0 |
| 0.844 | 6,250,000 | 1.0 |

| Search Time (secs) | Solution Cost | Optimality Bound ($\epsilon$) |
|---|---|---|
| 0.70 | 1,200,000 | 10.0 |
| 9.10 | 650,000 | 2.0 |
| 20.10 | 550,000 | 1.0 |

Middle table is for "walk with jumps over three river" example. The first solution is suboptimal—the character makes inefficient two legged jumps to cross all three rivers(Figure 2.6). The second solution is more natural, the character now uses a one-legged jump to cross the rivers(Figure 2.6). In the optimal solution the character does not jump but steps over the last (the smallest) river. In the bottom table, the character starts on the small rectangle and needs to walk toward and pick a small object shown by a sphere in Figure 2.5. The first solution is very suboptimal, the character bends way too far to pick up a small object (Figure 2.5). The second solution is better but the character is reaching from the side which in the absence of constraints appears unnatural(Figure 2.5). The final solution is optimal and looks natural (Figure 2.5).

## 2.7  Optimal Search

We use $A^*$ search[14], and in particular its anytime extension $ARA^*$ [13], to find the paths through the motion graph. We briefly describe it in Section 2.7.1. In Section 2.7.2 we analyze complexity of our problem and show why it is hard. To make optimal search tractable we propose a lossless compression of the motion graph that significantly reduced the number of states (Section 2.7.3) and a search heuristic that worked well for many examples of human motion(Section 2.7.4).

### 2.7.1  Search Method

We use $A^*$ search [14], and in particular its anytime extension $ARA^*$ [13], to find the paths through the motion graph and interpolation weights so that the interpolated path will satisfy the constraints and result in the optimal motion. The algorithm takes as input a graph where each edge has a strictly positive cost, a start state, $s_{start}$, and a goal state, $s_{goal}$. It then searches the graph for a path that minimizes the cumulative cost of the transitions in the path. $A^*$ uses a problem-specific heuristic function to focus its search on the states that are more likely to appear on the optimal path because they have low estimated cost. For each state $s$ in the graph, the heuristic function must return a non-negative value, $h(s)$, that estimates the cost of a path from $s$ to $s_{goal}$. To guarantee the optimality of the solution and to ensure that each state is expanded only once, the heuristic function must satisfy the triangle inequality: for any pair of states $s, s'$ such that $s'$ is a successor of $s$, $h(s) \le c(s, s') + h(s')$, where $c(s, s')$ is the cost of a transition between states $s$ and $s'$. For $s = s_{goal}$, $h(s) = 0$. In most cases, if the heuristic function is admissible (i.e., does not overestimate the minimum distance to the goal), the triangle inequality holds. For a given graph and heuristic function, $A^*$ searches the minimum number of states required to guarantee the optimality of a solution [16].

The anytime extension of $A^*$, $ARA^*$ search [13], trades off the quality of the solution for search time by using an inflated heuristic ($h$-values multiplied by $\epsilon > 1$). The inflated heuristic often results in a speedup of several orders of magnitude. The solution is no longer optimal, but its cost is bounded from above by $\epsilon$ times the cost of an optimal solution. $ARA^*$ starts by finding a suboptimal solution quickly using a large $\epsilon$ and then gradually decreases $\epsilon$ (reusing previous search results) until it runs out of time or finds a provably optimal solution.

### 2.7.2  Complexity

The complexity of the $A^*$ algorithm is $O(E + S log S)$, where $S$ is the number of states and $E$ is the number of edges in the graph. If a motion graph, $MG$ contains $10,000$ states, the unrolled graph $MG$ (without interpolation) will contain $S = 10^{12}$ if we discretize $P_x$ and $P_z$ into 1000 by 1000 values and $Q_{yaw}$ into 100 values. This graph cannot be searched quickly for an optimal solution. As a result, all existing approaches in the literature either find a solution using a global but sub-optimal approach with no guarantee on sub-optimality or search a manually constructed graph with a small number of states. The unrolled, interpolated motion graph, $ISG$, is even more challenging to search because it has a larger number of states ($S = 10^{17}$ for this example assuming we discretize $w$ into 10 values). Constraining the character to stay inside the corridor around the user-specified path would reduce the number of states to $S = 10^{15}$ if approximately 1% of the position values fall within the corridor. This reduction is not enough to make optimal search possible.

To address this problem, we developed two techniques that significantly decrease the number of states that the search will need to visit. The first technique compresses the motion graph into a practically equivalent but much smaller graph. The second technique computes an informative heuristic function that guides the search toward states that are more likely to appear in an optimal solution. In Section 2.8, we show that the combination of these techniques makes it possible to find an optimal or a close-to-optimal solution for a database of a reasonable size with a few minutes of computation. The next two sections give the details of both techniques.

### 2.7.3   Graph Compression

We compress the motion graph in two steps. First, we cull states and transitions that are sub-optimal. These states will not appear in the optimal solution for any set of user-specified constraints because the graph contains a lower cost alternative. Second, we cull states and transitions that are redundant because they are similar to other states and transitions in the motion graph. These steps result in a compressed version of graph $MG$ and the graph $IMG$ is derived from that graph as described in Section 2.7.3.

**Culling sub-optimal states and transitions:** To cull transitions, we first identify a specific class of states: those in which a contact change occurs (from double support to right leg contact, for example, or from no object in hand to object in hand). More formally, state $S$ in motion $M$ is defined as a *contact change state* if the state that directly precedes state $S$ in motion $M$ has a different set of contacts with the environment. Contacts are assumed to be not moving with respect to the ground plane or object in the environment. To determine the contact change states, we separate motions into phases based on contact with the environment. We use the technique of Lee and his colleagues [10] to identify the contacts and then verify them by a visual inspection (only a very small percentage of the contacts need to be adjusted by hand for locomotion and other simple behaviors). Contact information could also be computed using one of the other published techniques [3, 6].

We can then compute paths that connect pairs of contact change states without passing through another contact change state. All states in each such path will have the same contacts except for the terminal state where the contact changes. We call these paths *single contact* paths.

We can remove a large number of single contact paths from the graph $MG$. The key insight behind our algorithm is that although there are likely to be many single contact paths that connect two contact change states (thousands in our experiment), they all end with the character in exactly the same pose and with the same root position and orientation (Figure 2.7). Only one of these paths is optimal with respect to our optimization function. Therefore we can cull all other paths before unrolling the graph into the environment without reducing the functionality of the graph. Figure 2.8 illustrates this process.

This culling step does not affect the functionality of the graph unless the constraints provided by the user require controlling the details of the motion during a period of time when the contacts are not changing. For example, the

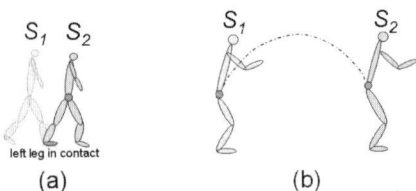

**Fig. 2.7.** (a) For direct paths between a pair of contact change states $S_1$ and $S_2$, the global position and orientation of the root of the character at state $S_2$ is uniquely determined by the contact position and orientation at state $S_1$ and the values of the joint angles at state $S_2$. (b) The position of the center of mass at landing (state $S_2$) is uniquely defined by the intersection of the flight trajectory and the center of mass of the character at state $S_2$. The trajectory of the center of mass for the root of the character is defined by the lift-off velocity from state $S_1$.

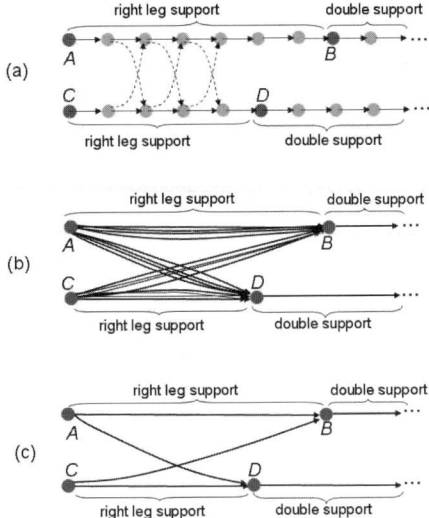

**Fig. 2.8.** (a) States $A$, $B$, $C$ and $D$ (shown in red) are contact change states. If the character enters state $A$ (and initiates a right leg support phase), it can exit only through state $B$ or $D$. (b) A representation with only the contact change states shows that there are many paths between each state. (c) The graph after transitions are culled to include only optimal paths between contact states.

user could no longer ask for waving while standing in place. Because animated characters tend to act on their environment, user constraints often create contact changes we have not found this restriction to be a serious problem. We can revert to searching an uncompressed motion graph if a constraint falls in the middle of the contact phase.

The optimal path might also violate environmental constraints if the swept volume for the character from one contact change state to the next intersects an obstacle. If the original graph contains a different path that would not have violated the constraints, then the culled graph will have lost functionality. This situation is uncommon because neither endpoint intersects the obstacle (or the search would not have explored the state) and only limited movement is possible with one contact change.

The optimization function we use allows us to compute optimal paths as a precomputation step because it is independent of the particular constraints the user specifies. Many common optimization functions are independent of the particular problem specification: minimizing energy, minimizing sum of squared accelerations, maximizing smoothness, minimizing the distance traversed, minimizing the total time of the motion, and satisfying specified annotations (for example behaviors or styles as in [2]). We can also support objective functions that depend on the user specification at contact change states. These functions can often be used to approximate other functions. For example, instead of minimizing the distance between every frame of the motion and the user-specified sketch, we can minimize the distance between the contact change states and the sketch.

Culling transitions in this way is different from retaining only the transitions between contact change states, as others have done [10]. With that approach no path would be found between states $A$ and $D$ in Figure 2.8(a) even though one exists in the original motion graph. The preprocessing presented here retains many more unique transitions, a property that is important for finding transitions between different behaviors such as a walk and a jump.

**Culling redundant states and transitions:** After we cull the sub-optimal states and transitions, we cull redundant ones. Motion graphs often include redundant data because of the need to capture natural transitions between behaviors. For example, to include natural transitions between walking and jumping, we included many steps of similar walking segments. As a result, each state in the motion graph may have many outgoing transitions that are similar. If we remove this redundancy, we can significantly reduce the size of the graph. This compression is performed on the graph that contains only contact change states, and the transitions are the optimal sequences of poses between contact change states.

For example, state $A$ in Figure 2.9 has three successors, $S_1$, $S_2$ and $S_3$, that are similar to each other and all three transitions will end at approximately the same position in the environment when the graph is unrolled. We can cull the redundant states by merging states $S_1$, $S_2$ and $S_3$ into one.

When two or more states are merged to form a new state, the successors of that state are the union of the successors of the merged states. Similarly, the predecessors are the union of the predecessors of the merged states. After merging, we will have many redundant transitions. We keep only the lowest cost transition (Figure 2.9). We merge states in the order of their similarity. The similarity threshold for merging can be substantially larger than that for

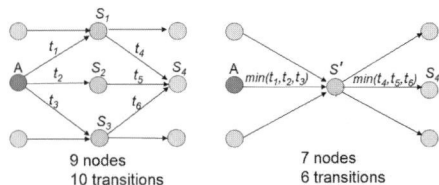

**Fig. 2.9.** (Left) States $S_1$, $S_2$ and $S_3$ are similar to each other. As a result, optimal transitions $t_1$, $t_2$ and $t_3$ are also very similar and all end with the character at approximately the same position. (Right) We merge states $S_1$, $S_2$ and $S_3$ into one state $S'$ and keep only the lowest cost transition.

establishing the initial transitions. A higher threshold for merging just removes flexibility from the graph whereas a higher threshold for transitions introduces perceptible errors. Because each transition in the compressed graph is a sequence of poses representing one contact phase of the motion, we can post-process each transition to remove foot-sliding as a preprocessing step.

**Constructing Compressed $IMG$:** Previous two section described how to compute a compressed standard motion graph. Comptressed interpolated motion graph can be computed from it. After graph $MG$ is compressed, graph $IMG$ is constructed from it using the same process as described in Section 2.5.2. In the compressed graph $MG$, however, each transition is a sequence of poses in between two contact change states. Consequently, a transition from state $A = (I_1^A, I_2^A, w_1^A)$ to state $B = (I_1^B, I_2^B, w_1^B)$ in graph $IMG$ is now a sequence of poses where each pose is an interpolation of corresponding poses in the transitions from $I_1^A$ to $I_1^B$ and from $I_2^A$ to $I_2^B$ in the compressed $MG$ (see [18] for more information). The interpolation weight $w$ is constant throughout the transition. We use the interpolation scheme described in Safonova and Hodgins [17]. When the durations of the transitions from $I_1^A$ to $I_1^B$ and from $I_2^A$ to $I_2^B$ differ, we assume a uniform time scaling with the time of the interpolated segment computed according to formula in [17].

As was shown by Safonova and Hodgins [17], this interpolation scheme ensures that the majority of the created transitions in graph $IMG$ are close to physically correct. We can also check these interpolated transitions for physical correctness using inverse dynamics at the time of construction of graph $IMG$.

### 2.7.4   Heuristic Function

We use an anytime version of the $A^*$ search algorithm to find an optimal path in the unrolled graph, $SG$. The number of states that $A^*$ search explores depends on the quality of the heuristic function—the lower bounds on cost-to-goal values. Informative lower bounds can significantly reduce the amount of the search space that is explored. In this section, we present a method for computing such bounds.

In Section 2.8 we show that this heuristic function usually speeds up the search by several orders of magnitude and is often the determining factor in whether a solution can be found.

The heuristic function must estimate the cost of getting to the goal while satisfying user and environmental constraints for each state $S$ in the graph $SG$. We compute two heuristic functions: $H_{pos}$ and $H_{mg}$. The first heuristic function, $H_{pos}$, ignores the dynamics of the motion of the character and estimates the cost of getting to the goal based only on the current position of the character, sketch of the user path and obstacles in the environment. The second heuristic function, $H_{mg}$, takes into account the capabilities of the character that are encoded in the motion graph but ignores its position in the environment. The combination of the two heuristics creates an informative measure of the cost of solving the problem specification. We now describe how to compute $H_{pos}$ and $H_{mg}$ and how to combine them. Same heuristic function can be used to search both standard motion graphs, $SMG$, and interpolated motion graphs, $ISG$.

**Heuristic based on the character location ($H_{\mathrm{pos}}$):** $H_{pos}(S, G)$ is the shortest path on the ground plane from the position of state $S$ to the position of the goal state $G$. The path must avoid obstacles and remain inside the corridor around the user-specified path (Figure 2.10). To compute $H_{pos}(S, G)$, we discretize the environment into 0.2 by 0.2 meter cells and compute the shortest path from the center of each cell to the goal. A single Dijkstra's search on a $2D$ grid can be used to compute all the paths with only a few milliseconds of computation. Because our cost function minimizes weighted average of energy and smoothness terms, we need to multiply the shortest distance (in meters) by an estimate of the minimum value in objective function required to traverse one meter. We compute the minimum value in objective function from the motion graph data. Because the computation of the heuristic $H_{pos}(S, G)$ depends on a given user sketch, it must be computed at runtime.

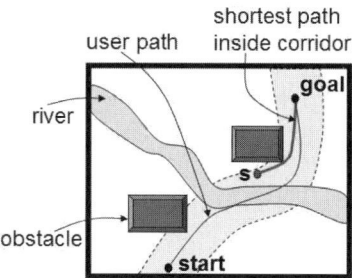

**Fig. 2.10.** $H_{\mathrm{pos}}(S, G)$ is the shortest path from the position of the character at state $S$ to the goal. The shortest path is constrained to stay inside the corridor.

We use a coarse discretization to compute the heuristic function, but a much finer discretization when computing the unrolled graph $SG$. For that computation, the root position of the character was discretized into a 0.05 by 0.05 meter grid to avoid discontinuities in the final motion.

**Heuristic based on motion graph state ($H_{mg}$):** $H_{pos}(S, G)$ provides a reasonable estimate of the cost to the goal for motions that simply require the character to travel from one location in the environment to another. But if constraints are present then $H_{pos}(S, G)$ will underestimate the cost to the goal for two reasons: (1) user or environmental constraints usually require much more effort than the minimum torque estimate assumed by $H_{pos}$; (2) the motion graph restricts what the character can do from a particular state, perhaps making a state that satisfies the constraint hard to reach. For example, if the character needs to jump over an obstacle and it is difficult to reach a jumping motion from state $S$, then the cost-to-goal at state $S$ should be high. $H_{pos}$ will grossly underestimate this cost and, consequently, $A^*$ search will needlessly explore this part of the space. The second heuristic function, $H_{mg}$, addresses this problem by taking into account the capabilities of the character that are encoded in the motion graph. It estimates the extra cost (the cost not accounted by $H_{pos}$) of satisfying each type of constraint for each state in the motion graph.

In our implementation, we support five types of constraints: picking, jumping, stepping onto an obstacle (a beam for example), ducking, and sitting. The method should generalize to other types of constraints such as kicking, stepping over obstacles, or standing on one leg. For each type of a constraint supported by our system, $H_{mg}(S, C)$ is computed as the minimum cost of getting to any pose in the motion graph that satisfies the constraint $C$ from the state $S$.

The computation of the $H_{mg}$ heuristic does not depend on the particular constraint specified by the user and therefore can be precomputed. The computation of the $H_{mg}$ heuristic is automatic because we have contact information for all motions in the motion capture database. We use the constraint of picking up an object to explain the computation of $H_{mg}$. The same method is used for all constraints supported by our system.

Both $H_{pos}(S, G)$ and $H_{mg}(S, C)$ account for the motion of the character in the plane. Therefore, the summation may overestimate the actual cost to the goal and violate the admissability requirement for the heuristic function. To resolve this problem, when computing the $H_{mg}(S, C)$ term, the cost of each transition in the motion graph is reduced by the minimum torque required to traverse the planar distance covered by the transition ($H_{pos} for that transition$). We denote this heuristic by $\tilde{H}_{mg}(S, C)$.

When the constraint, $C$, is to pick up an object, $H_{mg}(S, C)$ represents the minimal cost of a path in the motion graph from state $S$ to any state that represents the picking up of an object. Each "picking" pose can be defined by two parameters: *height* and *reach* (Figure 2.11(a)). *Height* is the height of the object with respect to the ground. *Reach* is how far the character must reach out to pick up the object (distance between the root and the hand projected onto the ground). Based on the contact information, we automatically identify each state

state $S_i$

| height / length | 0.2-0.4 meters | 0.4-0.6 meters | 0.8-1.0 meters | 1.0-1.2 meters |
|---|---|---|---|---|
| 0.2-0.4 meters | cost | cost | cost | cost |
| 0.4-0.6 meters | cost | cost | cost | cost |
| 0.8-1.0 meters | cost | cost | cost | cost |
| 1.0-1.2 meters | cost | cost | cost | cost |

(a)                              (b)

**Fig. 2.11.** (a) We identify states where objects are picked up in the motion graph. Each such pose is parameterized by two parameters: *height* and *reach*. (b) For each state in the motion graph we precompute a table with the minimal cost of getting to a "picking" state with the specified height and reach parameters.

in the graph $MG$ that represents picking up an object and compute *height* and *reach* values for that state. For graph $IMG$, the process is similar. We identify each state $p = (I_1, I_2, w)$ that represents picking up an object. Both poses, $I_1$ and $I_2$ are states where an object was picked up. At this state, the character assumes a picking pose with *height* and *reach* values based on the interpolation of poses $I_1$ and $I_2$ with weight $w$.

For each state in the graph $MG$(process is the same for $IMG$), we then compute a table (Figure 2.11(b)) where each cell represents a range of *height* and *reach* values, and the value is the minimal cost of getting from the given state to a state that represents picking with *height* and *reach* values in this range. For each entry in the table and each state in $MG$, we search graph $MG$ to compute the cost for that entry. The computation is really fast for graph $MG$. For graph $IMG$, for the database of 6-7 minutes of motion the precomputation of the $H_{mg}$ heuristic for all constraints took less than an hour.

**Combining the two heuristics:** We combine $H_{\mathrm{pos}}(S, G)$ and $H_{mg}(S, C)$ into a single heuristic function by summing them together. If at state $S$ there are still $n$ constraints remaining to be satisfied, we fetch the $H_{mg}$ term for each of these constraints, and then add all of them to the $H_{\mathrm{pos}}(S, G)$ term to obtain a heuristic value for state $S$:

$$H(S) = H_{\mathrm{pos}}(S, G) + \sum_{i=1...n} H_{mg}(S, C_i) \qquad (2.2)$$

## 2.8   Experimental Results

To illustrate the effectiveness of our approach, we generated a variety of examples for both standard motion graphs ($MG$) and interpolated motion graphs ($IMG$). For each experiment, the user specified a 2D path in the environment that the character should follow and the width of the corridor around that path. In some

**Fig. 2.12.** A forty-five meter motion with jumps and walks

experiments, the user also specified constraints such as picking up an object or sitting on a chair. Based on the sketch and the current environment, the system automatically computed environmental constraints such as stepping onto an obstacle, ducking, and jumping.

### 2.8.1    Search Effectiveness for Standard Motion Graphs (MG)

We have generated a variety of different examples including various walks such as straight walks and walks with slow and sharp turns, stepping over stones and jumps of different lengths and of different types (one-legged jumps and two-legged jumps). The figure 2.12 shows a screenshot of one of the motions generated by our algorithm.

For all the experiments we used a motion database containing approximately 12,000 frames of human motion captured at 30 frames per second. The motions included various walks, jumps, and picking up objects motions. The motion database was used to generate a single motion graph for all the experiments. The generated motion graph was relatively densely connected with the number of states being the same as the number of frames and the number of edges equal to about 250,000. While the density of the transitions in general makes the task of the search harder, it has a significant benefit in that we can generate motions that are quite different from the motions in the motion database. This freedom allows motions to follow the user specified path and meet the constraints well. In addition, because the generated solutions minimized the sum of squared accelerations (an approximation to energy), the motions generally avoided various artifacts such as dithering back and forth motions. All the generated motions were post-processed to remove feet sliding, an effect usually seen in motions that come straight from a motion graph. (The post-processing used a very simple local optimizer.)

**Table 2.2.** The cost of a solution and the bound on its sub-optimality as a function of search time. (The numbers are only given for the time points when the solution cost changed.)

| Search Time (secs) | Solution Cost | Optimality Bound ($\epsilon$) |
|---|---|---|
| 0.016 | 10658960 | 10.0 |
| 0.031 | 10658960 | 2.1 |
| 0.032 | 8226729 | 2.0 |
| 0.033 | 8226729 | 1.4 |
| 0.250 | 6268591 | 1.3 |
| 0.844 | 6268591 | 1.0 |

The lengths of the motions generated by our approach varied from 8 to 60 seconds. The longest one was a 45 meter walk through a maze that also involved a number of examples of jumping over water. In all the examples the search quickly generates a first solution (for sub-optimality bound $\epsilon$ set to 10): usually within 2 to 3 seconds. In case of the long walk through the maze though the first solution was generated within 30 seconds. The search then improves the solution within the remaining time allocated to it. For each example the search was given 120 seconds to find the best solution it can. The table 2.2 shows how the cost of the found solution improves during the search for one of the simpler experiments. $\epsilon$ set to 1 corresponds to a provably optimal solution. Usually, the search converged to somewhere in between $\epsilon = 1.1$ (at most, 10% sub-optimality) and $\epsilon = 1.5$ within the allocated time.

### 2.8.2 Search Effectiveness for Interpolated Motion Graphs (IMG)

Figure 2.1 shows the motion of a character traversing an obstacle course. The character walks over the beam, jumps over holes, ducks under a bar, and finally sits on a chair. This example illustrates that our algorithm can synthesize motions that are 15 seconds long and consist of several different behaviors. Besides the obstacle course examples, we have also synthesized many other examples, including walking along paths of varying curvature, picking and placing an object in various locations, jumping over stones with variable spacing, jumping with different amounts of rotation and distance, and forward walks of different step lengths. Figure 2.13 shows images for some of the results. For shorter, single behavior examples, such as jumps and short walks, only a few milliseconds to a few seconds were required to compute an optimal solution. For longer, multi-behavior examples, a few minutes were required to compute a close-to-optimal solution. In general, the time depends on the size of the database, the length of the generated motion, and the complexity of the constraints.

### 2.8.3 The Benefit of Motion Graph Compression

In this experiment, we evaluate the effect of motion graph compression. Table 2.3 shows these statistics for three different databases: (1) walking, jumping, ducking, sitting and walking along a beam; (2) walking and picking up an object;

**Fig. 2.13.** Synthesized motions

**Table 2.3.** Compression for three motion graphs. The first graph is computed from walking, jumping, ducking, sitting and walking along the beam motions. The second graph is computed from walking and picking motions and the third one is computed from just walking motions.

|       | Before merging | After removing sub-optimal data | After removing redundant data | Compression time |
|-------|----------------|---------------------------------|-------------------------------|------------------|
| DB 1  | states=6,000 trans=90,000 | states=350, trans=12,500 | states=130 trans=700 | 30 min |
| DB 2  | states=12,000 trans=250,000 | states=700 trans=60,000 | states=300 trans=5,000 | 60 min |
| DB 3  | states=2,000 trans=25,000 | states=173 trans=3,700 | states=50 trans=300 | 2 min |

(3) just walking motions. For each database, we computed the number of states and transitions in the motion graph before compression, after the first compression step (removing sub-optimal data), and after the second compression step (removing redundant data). The table also gives the time required to compress the graph (a precomputation step performed only once for each database). Compression techniques reduce the size of the graph by a factor of 20 to 50.

### 2.8.4  The Benefit of the Heuristic Function

We also evaluated the effectiveness of our heuristic function. The results are shown in Table 2.4. We compare four heuristics: (1) the Euclidean distance to the goal; (2) the $H_{pos}$ component of our objective function; (3) the $H_{mg}$ component of our objective function; (4) the combined heuristic function with both $H_{pos}$ and $H_{mg}$ components. The results demonstrate that our heuristic function is essential for making the search efficient and often makes the difference between

**Table 2.4.** Evaluation of the heuristic function for the problem of picking up an object. We sampled the location of the object into 179 samples. Each column shows the average search time in seconds, the average number of states expanded during the search and the percent of the experiments that succeeded (found solution within 10 minutes and did not run out of memory). The statistics are reported for the Euclidean distance to the goal, the $H_{2D}$ component alone, the $H_{mg}$ component alone, and the combined heuristic function. The first row shows results for a solution whose cost is at most 10 times the optimal one. The sub-optimality bound for the second row is 3 and the solution in the last row is optimal.

| $\varepsilon$ | Euclidean distance | | | $H_{2D}$ | | | $H_{mg}$ | | | $H_{combined}$ | | |
|---|---|---|---|---|---|---|---|---|---|---|---|---|
| | time | exp | solved | time | exp | solved | time | exp | solved | time | exp | solved |
| 10.0 | 8.0 | 185,813 | 100% | 8.1 | 160,718 | 100% | 11.6 | 72,004 | 100% | 0.8 | 9,332 | 100% |
| 3.0 | 17.1 | 481,321 | 100% | 16.8 | 406,149 | 100% | 15.1 | 103,000 | 100% | 1.6 | 16,068 | 100% |
| 1.0 | 100.2 | 1,832,347 | 20% | 97.8 | 1,748,620 | 20% | 48.1 | 270,812 | 80% | 49.5 | 275,712 | 80% |

finding a good solution and not finding one at all. The table also shows that both components of the heuristic function are important, neither component alone is effective.

## 2.9 Discussion

Motion graphs and their variations have proven to be a a powerful technique to solve for a desired motion when only rough sketch is given. In this paper, we have demonstrated that it is possible to search a standard motion graph and interpolated motion graph using a globally optimal search algorithm, $A^*$. Two contributions made this possible: a lossless compression of the motion graph that significantly reduced the number of states and a search heuristic that worked well for many examples of human motion. We demonstrated that the global search was effective by creating long example motions and showing that the optimal and near-optimal solutions avoided the dithering and inefficient patterns of motion seen in many other motion graph implementations.

Because the method computes a compressed motion graph that contains only optimal paths, variations that may have existed in the original data may be lost. Variations are always "sub-optimal" and therefore will be culled. We would like to experiment with keeping several maximally different paths rather than just one. In our experience, most of what is culled is redundant trajectories that are visually indistinguishable but additional experiments would be required to decide whether important variability is lost.

The quality of the results largely depend on the quality of the motion database used to construct the motion graph. For example, if the database contains only a motion of sitting on a tall chair then we cannot synthesize a motion for sitting on a medium or a low height chair because there are no two motions whose interpolation would provide the desired motion. We also found that the motion graph must have "good" connectivity. Our experiments show that to obtain good results many states must be able to quickly connect to the constraint states and vice versa.

Global optimization has two significant effects on the motion of the character. First, it should iteratively find the "correct" strategy for the character to use to navigate an environment. For example, is a two-legged jump or a one-legged jump more efficient for an obstacle of a given size? Table 2.1 illustrates these discrete changes. The second feature of the global optimization should be to fine tune the motion, choosing a series of walking steps with little velocity change, for example. This second feature is not as apparent in the animated motion but is still visible in the decrease in energy as the optimizer iterates.

Although we did a few informal experiments to see if a subject used the same strategies as the animated character for a given terrain, we did not do a definitive assessment with a large naive subject pool. Such a study would be easy to run (using coding of the behavior selected for each part of the obstacle course as the metric). We expect that the sequence of behaviors from the human subjects would be similar to those of the animated character for many but not all examples. They might differ because the motion graph did not include the right behaviors (a long one-legged jump, for example) so another less efficient behavior is selected (a long two-legged jump, for example). Alternatively differences might arise because people do not always optimize efficiency but instead optimize for style, comfort, safety or other factors.

# References

1. Arikan, O., Forsyth, D.A.: Interactive motion generation from examples. ACM Trans. on Graphics 21(3), 483–490 (2002)
2. Arikan, O., Forsyth, D.A., O'Brien, J.F.: Motion synthesis from annotations. ACM Trans. on Graphics 22(3) (2003)
3. Callennec, B.L., Boulic, R.: Robust kinematic constraint detection for motion data. In: ACM SIGGRAPH/Eurographics Symp. on Comp. Animation, pp. 281–290 (2006)
4. Choi, M.G., Lee, J., Shin, S.Y.: Planning biped locomotion using motion capture data and probabilistic roadmaps. ACM Trans. on Graphics 22(2), 182–203 (2003)
5. Fang, A.C., Pollard, N.S.: Efficient synthesis of physically valid human motion. ACM Trans. on Graphics 22(3), 417–426 (2003)
6. Ikemoto, L., Arikan, O., Forsyth, D.: Knowing when to put your foot down. In: ACM Symposium on Interactive 3D Graphics, pp. 49–53 (2006)
7. Kovar, L., Gleicher, M., Pighin, F.: Motion graphs. ACM Trans. on Graphics 21(3), 473–482 (2002)
8. Lau, M., Kuffner, J.: Precomputed search trees: Planning for interactive goal-driven animation. In: ACM SIGGRAPH/Eurographics Symp. on Comp. Animation, pp. 299–308 (September 2006)
9. Lau, M., Kuffner, J.J.: Behavior planning for character animation. In: ACM SIGGRAPH/Eurographics Symp. on Comp. Animation, pp. 271–280 (2005)
10. Lee, J., Chai, J., Reitsma, P.S.A., Hodgins, J.K., Pollard, N.S.: Interactive control of avatars animated with human motion data. ACM Trans. on Graphics 21(3), 491–500 (2002)
11. Lee, J., Lee, K.H.: Precomputing avatar behavior from human motion data. In: ACM SIGGRAPH/Eurographics Symp. on Comp. Animation, pp. 79–87 (2004)

12. Li, Y., Wang, T., Shum, H.-Y.: Motion texture: a two-level statistical model for character motion synthesis. ACM Trans. on Graphics 21(3), 465–472 (2002)
13. Likhachev, M., Gordon, G., Thrun, S.: ARA*: Anytime A* with provable bounds on sub-optimality. In: Advances in Neural Information Processing Systems (NIPS), vol. 16. MIT Press, Cambridge (2003)
14. Pearl, J.: Heuristics: Intelligent Search Strategies for Computer Problem Solving. Addison-Wesley, Reading (1984)
15. Pullen, K., Bregler, C.: Motion capture assisted animation: texturing and synthesis. ACM Trans. on Graphics 22(2), 501–508 (2002)
16. Russell, S., Norvig, P.: Artificial Intelligence: A Modern Approach. Prentice-Hall, Englewood Cliffs (2003)
17. Safonova, A., Hodgins, J.K.: Analyzing the physical correctness of interpolated human motion. In: ACM SIGGRAPH/Eurographics Symp. on Comp. Animation, pp. 171–180 (2005)
18. Safonova, A., Hodgins, J.K.: Construction and optimal search of interpolated motion graphs. ACM Trans. Graph., 106 (2007)
19. Safonova, A., Hodgins, J.K., Pollard, N.S.: Synthesizing physically realistic human motion in low-dimensional, behavior-specific spaces. ACM Trans. on Graphics 23(3), 514–521 (2004)
20. Sulejmanpašić, A., Popović, J.: Adaptation of performed ballistic motion. ACM Trans. on Graphics 24(1), 165–179 (2005)
21. Sung, M., Kovar, L., Gleicher, M.: Fast and accurate goal-directed motion synthesis for crowds. In: ACM SIGGRAPH/Eurographics Symp. on Comp. Animation, pp. 291–300 (July 2005)
22. Witkin, A., Kass, M.: Spacetime constraints. Computer Graphics (Proceedings of SIGGRAPH 88) 22(4), 159–168 (1988)

# 3

# Motion Synthesis with Adaptation and Path Fitting

Newman Lau, Chapmann Chow, Bartholomew Iu, and Pouro Lee

Multimedia Innovation Center, School of Design, The Hong Kong Polytechnic University

**Abstract.** In this paper, we present an algorithm for motion synthesis. We develop this algorithm as a part of a large crowd simulation project which simulates the motion of the people by using a large database of motion captured data. Two main features of this synthesis algorithm include generating motion along a given path with the speed matched and adapting the motion to the scaled skeleton. In some previous works, the similar algorithm are offline algorithm, memory consuming and possibly need to search neighbor time frames back and forth. In order to develop an online algorithm to be used in the large crowd simulation with small memory footprint, we develop this alternative solution.

## 3.1 Introduction

In this paper, we present an algorithm for motion synthesis. This algorithm is developed as a part of a large crowd simulation project which simulates the motion of the people by using a large database of motion captured data. Most of the crowd simulations divide the simulation process into two phases, the first phase is path generation, and the second phase is matching the motion clip to the generated path. The second phase is the goal of this paper. There are two problems we need to solve for the second phase. The first problem is that the generated motion should be attached to the path, and the speed of the generated motion must be matched with the path. The second problem is that the crowd simulation simulates different size of people, that is, the skeleton is scaled.

Because this algorithm is only a part of a large system, the memory is a scared resource. We, therefore, need to limit the memory usage. The best strategy to achieve this is to write the simulated frame data into hard disk and keep only the current simulating frame in the memory. Since the past data is not in memory, and the future data is not available, we cannot use the spacetime method [1, 2, 19]. We, therefore, develop our solution as an online algorithm. An online algorithm is not just suitable for our situation, but also suitable for game environment in which future data is not available.

The second method to limit the memory usage is to reduce the size of the motion database. We find that there are two possible strategies to reduce the size. The first one is to generate different speed walking motion using a single walking motion, and the second one is to generate small angle turning walking from a straight line walking motion. With these two strategies, we can greatly reduce the size of the motion database. There is one more benefit we gain from the first strategy. This benefit is that we don't need to specify the transition motion which represents the transition from fast walking to slow walking and vice versa.

D. Plemenos, G. Miaoulis (Eds.): Arti. Intel. Techn. for Comp. Graph., SCI 159, pp. 41–53.
springerlink.com                                    © Springer-Verlag Berlin Heidelberg 2009

In handling the motion captured data with a scaled destination skeleton, constraints such as the foot must be snap to a specific position at certain frames will be violated. In order to fulfill the constraint requirement, we must preserve the constraint while not destroy the quality of the motion captured data.

## 3.2 Related Works

There are some works focusing on generating a new motion which follows a predefined path from a database of motion clips [5, 9, 11]. The characteristics of these methods are that no time, speed and orientation requirement imposed on the predefined path, therefore, different generated motion may have different finishing time at the end of the path. Earlier approaches of simulating a motion with satisfying the time, speed and orientation requirement are already established. These approaches are named spacetime method [1, 2, 19]. However, there are some problems when used in practice, because many constraint must be set and some constraint are even not possible to specify, such as singing in the rain [2] and they takes a long computational time due to the fact that they use a non-linear solver.

Recently, there is a most similar method [17] to solve our problem. Their approach is to incorporate modification of a skeleton's position, orientation and speed in the motion graph searching process. A set of motion clips that is closely matched to the path are found, and then the motion clips are combined together to form a new motion. In contrast, our algorithm works on a small database of motion clip, and there are no different speed of walking motion, which is a reason why we can keep the motion clip database small. Another difference is the searching method, although this is not the focus of this paper, we mention here for a comparison. Our searching method need only the motion type as an input, hence we can reduce the possibility of finding a dead end clip from the motion graph [9]. Because the number of dead end clip can be reduced, we don't need to put transition clips in the motion database.

## 3.3 Requirement

We make some assumptions on the skeleton structure, motion clip and the path before proceed to the algorithm formulation. These assumptions have an influence in the choice of the axis used in the equations, and some assumptions are commonly used in animation software.

### 3.3.1 Skeleton Structure

We use the bvh format in our simulation. Actually, any format can be used. What we need is a unique structure of the skeleton's lower body. In this paper, we only focus on the constraint set on the foot, therefore, any upper body skeleton structure can be used. The structure of the left foot is: Root -> Left Hip -> Left Knee -> Left Ankle -> Left Toe. The right foot is similarly defined.

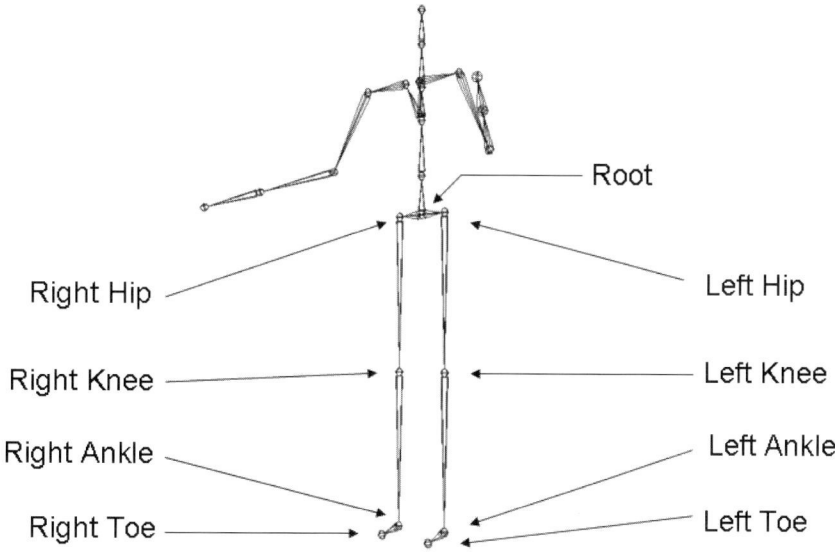

**Fig. 3.1.** Skeleton structure

### 3.3.2 Motion Clip

The motion clip we used in the simulation comes from the motion capture system. In order to reduce the size of the database of the motion clips, we have developed our algorithm with automatically turning feature. Therefore, the motion clip are all walking or running along a straight line. In case of a standing type motion, the actor always

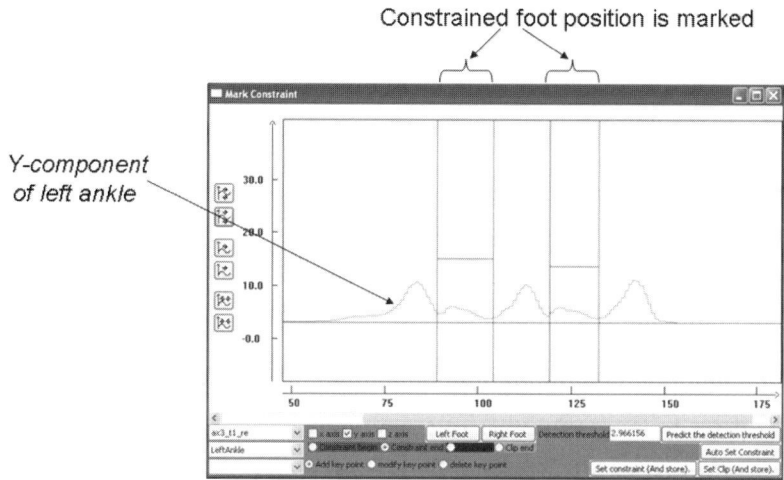

**Fig. 3.2.** Foot constraint setting

faces in the forward direction. The forward direction in our simulation is +Z axis, and the actor also walk in the +Z direction.

There are some other information must be stored with the clip before the computation. The first one is the type information, and the second one is the constrained position. In our current implementation, only two types of motion are allowed. One is walk type, which refers to walking and running. Another one is the stand type, which refers to motion with the actor staying around the origin position in the world.

The constrained position remembers when the foot is snapped to the ground. We mark the constrained position by using the algorithm mentioned in [7], and fine tune the constrained position by hand.

### 3.3.3 Path

In our simulation, we match the speed of the motion clip to that of the path. This path is actually the trajectory of the scene node. The path includes the position, orientation and scale information. With the scale information, we can simulate the motion of different size skeleton in the simulation.

We match the speed by attaching a skeleton center to the scene node. This skeleton center is not the root of the skeleton, but it is a rough center of the skeleton. The decision of using a skeleton center comes from a fact that the skeleton root, which is the hip, is not fixed in some stand type of motion, and the hip move with small vibration even in a straight line walking motion. If we plot the movement of the hip of a straight line walking motion, we can see that the hip move in a straight line with some ripples. Therefore, if we attach the skeleton root to the scene node directly, then the high frequency content of the hip movement will be lost, the generated motion looks odd and the foot skating problem is exaggerated. Another advantage of using the skeleton center is that we can extend the algorithm to including the stand-up and sit-down type of motion.

The only requirement of the path is a smooth derivative of the position and orientation. If the derivative of the orientation is not smooth, the generated motion may have a jittered limb. The smooth derivative of position is critical for a calculation step in the following sections. If the position derivative is not smooth, the generated motion looks jittered.

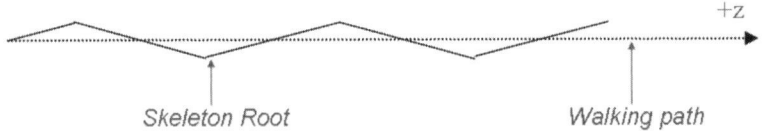

**Fig. 3.3.** Skeleton root moves with ripples along a straight path

## 3.4 Motion Clip Preprocessing

In this paper, there are two types of global position mentioned. One is skeleton global position which is calculated by the relative translation and rotation of the joints. Another one is the path global position which is the skeleton global position transformed by the path's translation, rotation and scale, and this transformation is represented by **T**.

In this preprocessing stage, we calculate four vectors from the motion clip for the later sections, they are:

- **fc**$_{i,j}$, the foot center, where subscript $i$ is the motion clip index, $j$ is the frame number.
- **rfc**$_{i,j}$, the relative foot center.
- **rfcd**$_{i,j}$, the root foot center difference.
- **jfcd**$_{i,j,k}$, the joint foot center difference, the subscript $i$, $j$ is similarly defined as **fc**$_{i,j}$, and $k$ is the index of joint of the skeleton.

Foot center **fc**$_{i,j}$ (Fig. 3.4) is the average skeleton global position of the left ankle and right ankle.

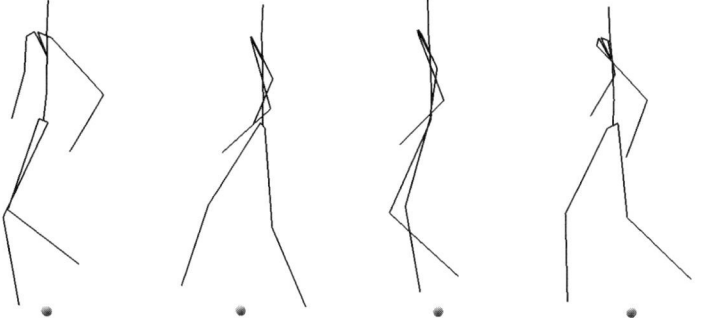

**Fig. 3.4.** The sphere represents the foot center

In the computation, we don't need the y position data, and we set it to zero. Notice that we use only the ankles for the calculation because ankles must be kept above the ground. An alternative calculation of skeleton center is the average position of all joints of the skeleton, but the result is worse if the motion is a sit-down or stand-up motion.

Relative foot center **rfc**$_{i,j}$ is calculated as **fc**$_{i,j}$ - **fc**$_{i,j-1}$, and **rfc**$_{i,j}$ is zero if $j$ is 0.

Root foot center difference **rfcd**$_{i,j}$ is calculated as the skeleton global position of skeleton root in frame $j$ minus the foot center **fc**$_{i,j}$.

Joint foot center difference **jfcd**$_{i,j,k}$ is calculated as the skeleton global position of the joint $k$ in frame $j$ minus the foot center **fc**$_{i,j}$.

## 3.5  Synthesis Algorithm

The synthesis algorithm can be divided into seven steps:

1. Skeleton center calculation.
2. Skeleton root calculation.
3. Joint space parameter processing.

4. Walking motion clip scaling.
5. Snap position shifting calculation.
6. Center pull.
7. Foot snap.

The first four steps are responsible for the path fitting part, and the rest of the steps are responsible for maintaining the foot constraints. We focus on the constraint set on the ankles only. For the toes, algorithm such as [10] can be applied to enforce the constraint set on toe.

Before going into the each step of the algorithm, we first describe the interpolation method used in our simulation and define some terms. For clarity, we describe how to process on the vector type data, which is not necessary the joint space parameter.

1. Find a motion clip $m_i$, copy the data of the form 0 to $length(m_i) - b - 1$ frames to the destination clip, where $b$ is the length of the blending interval, and $length()$ is a function to get the length of the input motion clip.
2. At the frame $length(m_i) - b$, set the clip $m_i$ to $m_{i-1}$, and find a new clip $m_i$.
3. In the blending interval, $length(m_i - 1) - b$ to $length(mi - 1) - 1$ of $m_{i-1}$, and 0 to $b - 1$ of $m_i$, we perform interpolation on the data. We use a cubic equation, which has a $C^1$ continuity, to blend the data. The cubic equation is $t = 2x^3 - 3x^2 + 1$, where $x$ is a floating number from 0 to 1 in this blending interval. A general blending equation is

$$\mathbf{c}_t = t\mathbf{a}_{fp} + s\mathbf{b}_{fc}, \ s = 1 - t, \tag{3.1}$$

where $\mathbf{c}_t$ is the result data of the generated clip of the target frame $t$; $\mathbf{a}_{fp}$ is the data in clip $m_{i-1}$, and $fp$ is the frame number in this range $length(m_{i-1}) - b$ to $length(m_{i-1}) - 1$; $\mathbf{b}_{fc}$ is the data in clip $mi$, and $fc$ is the frame number in this range 0 to $b - 1$.
4. For the frame from $b$ to $length(m_i) - 1$ of clip $m_i$, copy the data to the target skeleton and finish the algorithm. Repeat the processing from step 1 if there are frames need to be simulated.

For gesnerality, equation (3.1) is used for the frames outside the blending interval, and $t$ is always 0, $\mathbf{a}$ is undefined.

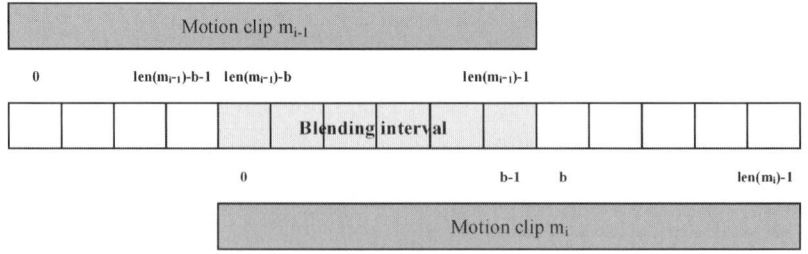

**Fig. 3.5.** Blending frames

### 3.5.1 Skeleton Center Calculation

Skeleton center, which is not the skeleton root, represents the rough center of the skeleton, and this center is used to attach the skeleton to the scene node.

The initial value of skeleton center is zero. Skeleton center $\mathbf{sc}_t$ of the frame $t$ is calculated as:

$$\mathbf{sc}_t = \mathbf{sc}_{t-1} + \partial_{i-1}t\mathbf{rfc}_{i-1,fp} + \partial_i s\mathbf{rfc}_{i,fc} \tag{3.2}$$

where $\partial_i$ is a checking variable, the z component will be zero if the clip $i$ is a walking motion, otherwise, the value is 1.

### 3.5.2 Skeleton Root Calculation

Skeleton root $\mathbf{sr}_t$ is calculated from the skeleton center. The root position keeps a short distance to the scene node even the motion clip is a walking motion. The skeleton root is calculated as:

$$\mathbf{sr}_t = \mathbf{sc}_t + t\mathbf{rfcd}_{i-1,fp} + s\mathbf{rfcd}_{i,fc} \tag{3.3}$$

### 3.5.3 Joint Space Parameter Processing

Joint space parameter is the joint's rotational data. The joint space parameter of each joint $\mathbf{q}_{t,j,k}$ of the destination skeleton is calculated as:

$$\mathbf{q}_{t,j,k} = t\mathbf{q}_{i-1,fp,k} + s\mathbf{q}_{i,fc,k} \tag{3.4}$$

where the subscript is similarly defined as $\mathbf{jfcd}_{i,j,k}$.

### 3.5.4 Walking Motion Clip Scaling

This scaling step is only applied to the motion clip that is the walk type, but for generality, we formulate the equation that applies to any clip type. The idea of the step is first calculate the original position of the ankle. Then scale the z component of the original position with respect to foot center. Finally we apply an IK algorithm to the foot with the scaled position as input.

The original position is calculated as:

$$\mathbf{p}_k = t\mathbf{jfcd}_{i-1,fp,k} + s\mathbf{jfcd}_{i,fc,k} \ . \tag{3.5}$$

In order to scale the position with respect to foot center, we add a scaling term $\partial_i$. The scaled position becomes

$$\mathbf{p'}_k = \partial_{i-1}t\mathbf{jfcd}_{i-1,fp,k} + \partial_i \, s\mathbf{jfcd}_{i,fc,k} \ , \tag{3.6}$$

where the z component of $\partial_i$ is the speed of the path divided by the skeleton root speed of the clip $m_i$ if the clip $i$ is a walking motion, otherwise, the value is 1.

### 3.5.5 Snap Position Shifting Calculation

Since the motion may be scaled in section 3.6.4, the constraint position will be shifted. In order to keep the foot snapping position, an error correction term should be added to the generated motion. This error correction term is then minimized bit by bit in the non-blending interval. The error correction term $\mathbf{shift}_k$ of joint $k$ will be set only at the first frame of the blending interval and joint $k$ is constrained in either clip $m_i$ or $m_{i-1}$. The error correction term $\mathbf{shift}_k$ is:

$$\mathbf{shift}_k = \mathbf{shift}_k + \mathbf{p}_k - \mathbf{p'}_k \qquad (3.7)$$

### 3.5.6  Center Pull

In combining several short clips together, the center shifting will be accumulated. If, for example, the skeleton's root of a short clip at the end frame has an offset (-0.1, 0, 0) with respect to that of the start frame, the accumulation shift of combining 100 same clips is (-10, 0, 0). Therefore, we need a center pull to keep the skeleton root close to the scene node. In order to simplify the calculation, the center pull will only be updated on non-blending interval. The amount of center pull is stored in the variable center pull accumulation $\mathbf{cpa}$ which is initially zero.

First we calculate a prediction variable:

$$\mathbf{pre} = \mathbf{sc}_t + \mathbf{cpa} \qquad (3.8)$$

and then set a target position:

$$\mathbf{tar} = \partial_i \mathbf{fc}_{i,fc} \qquad (3.9)$$

where $\partial_i$ is a checking variable, the z component will be zero if the clip $i$ is a walking motion, otherwise, the value is 1. The difference is:

$$\mathbf{diff} = \mathbf{pre} - \mathbf{tar} \qquad (3.10)$$

The update of $\mathbf{cpa}$ is done on each axis:

$$\mathbf{cpa}[a] = \mathbf{cpa}[a] + \partial\ thres, \ \ \text{if } abs(\mathbf{diff}[a]) > thres \qquad (3.11)$$

where $[a]$ take the axis $a$ data from the vector, $\partial$ is -1 if $\mathbf{diff}[a]$ is greater than zero, $abs()$ takes the absolute value of the input and $thres$ is a threshold value. The last step is to apply the center pull to the skeleton root:

$$\mathbf{sr}_t = \mathbf{sr}_t + \mathbf{cpa} \qquad (3.12)$$

### 3.5.7  Foot Snap

The idea of foot snap is simple. When the ankle touches the ground, calculate a path global position as a snap position. The ankle remains in this snap position if the constraint is set. Once the constraint is off, calculate the difference between the snap position and the path global position of the ankle. Another source of difference comes from the center pull in step 3.6.6. The difference will then be minimized bit by bit on the subsequence frames.

First we need to check the constraint transition. There are four cases: case 1, no-constraint; case 2, has-constraint; case 3, no-constraint to has-constraint and case 4, has-constraint to no-constraint. We represents the skeleton global position on frame $i$ of joint $k$ using $\mathbf{j}_{i,k}$.

Step 1, setting the snap position **snap**$_{i,k}$ on frame $i$, joint $k$. Different operation is applied on different cases:

- Case 1. Perform step A.
- Case 2. **snap**$_{i,k}$ = **snap**$_{i-1,k}$.
- Case 3. **snap**$_{i,k}$ = **T**(**j**$_{i,k}$ + **shift**$_k$).
- Case 4. **shift**$_k$ = **T**$^{-1}$ **snap**$_{i,k-1}$ − **j**$_{i,k}$, then perform step A.

Step A:

If |**shift**$_k$| is larger than a threshold, minimize **shift**$_k$, and set

$$\mathbf{snap}_{i,k} = \mathbf{T}(\mathbf{j}_{i,k} + \mathbf{shift}_k).$$

If |**shift**$_k$| is smaller than a threshold, unset **snap**$_{i,k}$.

Step 2, if **snap**$_{i,k}$ is set, apply IK on joint $k$ using **snap**$_{i,k}$ as the target.

## 3.6  Examples

In this section, we present some results made by our simulation.

### 3.6.1  Example 1

Fig. 3.6 shows a smooth motion generated by our simulation. The speed of the path is varying along the path, and the speed is within a reasonable range that the character does not need to make a long jump.

**Fig. 3.6.** The snapshots show a smooth motion generated by reasonable settings

### 3.6.2  Example 2

Fig. 3.7 shows that our simulation can generate a transition motion from walking to non forward walking. This non forward walking is not a stop motion, because the skeleton steps the feet on the same place. This situation happens if the walk type motion clip is too long, and the path is about to stop, then the simulation freeze the forward movement of the walk type motion clip. Handling in this way is better than suddenly stopping using the walk type motion clip and switch immediately to a stand type motion clip.

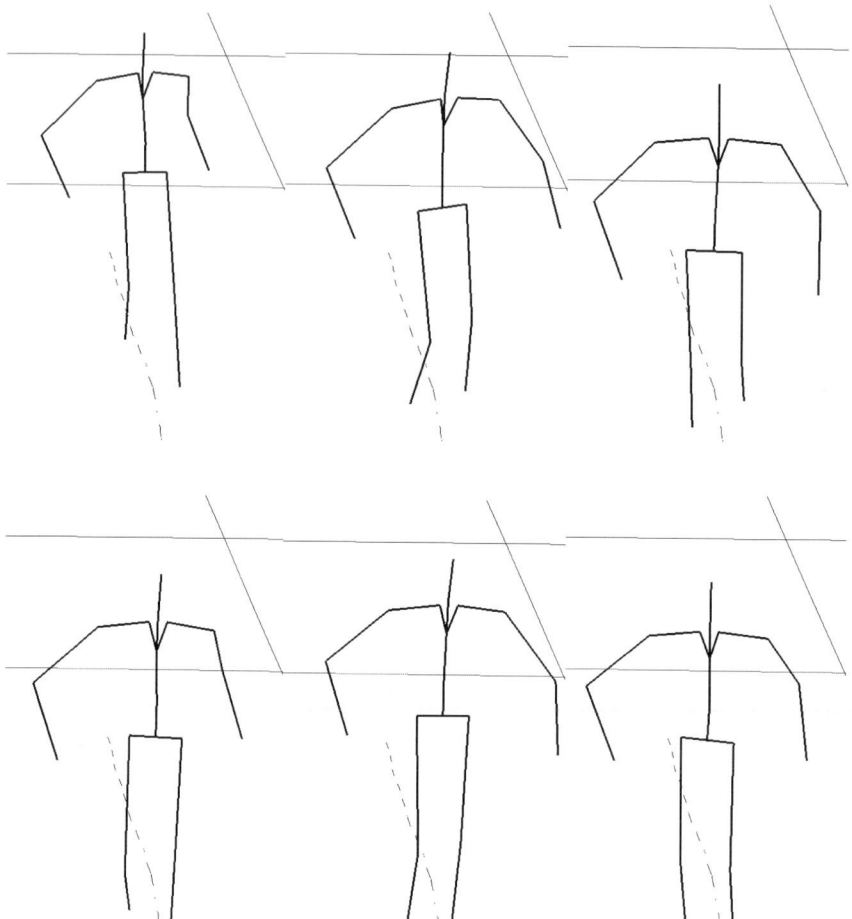

**Fig. 3.7.** The transition motion from walking to non forward walking is generated by walk type motion clip only

### 3.6.3  Example 3

Our algorithm does not handle the acute angle turning correctly, however, in some cases our algorithm can generate logically correct acute angle turning. In Fig. 3.8, the skeleton is performing a right turn with the right foot snapped to the ground, this turning is correct because the left foot does not kick on the right foot.

### 3.6.4  Example 4

Fig. 3.9 shows an unrealistic jump motion. This motion is generated by an unreasonable speed of the path, and this speed is much faster then that of the walk type motion clips. The motion graph used in this example contains no jump type motion. Although

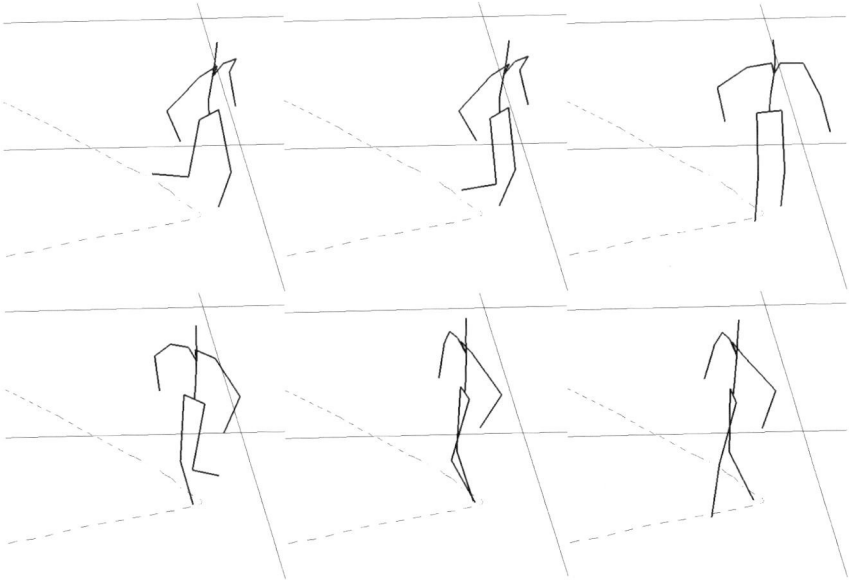

**Fig. 3.8.** An acute angle turning may be logically correct

**Fig. 3.9.** An unrealistic jump motion is generated by inappropriate setting

the motion is unrealistic, the motion is still a smooth motion and with the foot constraint enforced. This unrealistic example shows that our algorithm generates smooth motion even with non suitable input setting. Since our algorithm aims at generating motions for large crowd, unrealistic motion in a crowd is not noticeable, but a non smooth motion is more noticeable.

## 3.7  Discussion

In this paper, we presented an algorithm that generates a new motion clip from a database of motion clips. The new motion clip can have the skeleton scaled. While we maintain the foot constraint of the scaled skeleton, we also match the speed of the motion clip to that of a given path. There may be some small turning angles on the path, and our algorithm can handle these small turning angles nicely. If, however, an acute turning angle occurs on the path, our algorithm may generate a logically incorrect motion. What we mean by the logically incorrect is that the generated motion may have the left foot kicking on the right foot or vice versa. This problem can be solved by copying a motion clip that is performing an acute turning angle to the generated motion, and this is the future extension of the current implementation.

The next problem is that our current implementation only allows walk and stand type of motion, and the sit-down and stand-up motion type is not included. We know this problem when we design the algorithm. In order to allow for the extension, we introduce a skeleton center calculation in equation (3.2), and we attach the motion to the path using the skeleton center. With this skeleton center, the path generation can be simplified, because the path represents the foot pace, which is the skeleton center, but not the hip of the skeleton. In simulating a sit-down motion, when the path reach a chair, the path stops in front of the chair and the system play the sit-down motion. When the path is ready to leave the chair, the system plays a stand-up motion. If the motion clip is attached to the path using the hip, then the path need to reach the center of the chair, and the path generation is more complex.

## References

[1] Gleicher, M.: Motion editing with spacetime constraints. In: Proceedings of the 1997 symposium on Interactive 3D graphics, p. 139. ACM Press, New York (1997)

[2] Gleicher, M.: Retargetting motion to new characters. In: Proceedings of the 25th annual conference on Computer graphics and interactive techniques, pp. 33–42. ACM Press, New York (1998)

[3] Gleicher, M.: Motion path editing. In: Proceedings of the 2001 symposium on Interactive 3D graphics, pp. 195–202. ACM Press, New York (2001)

[4] Hasegawa, S., Toshiaki, I., Hashimoto, N., Salvati, M., Mitake, H., Koike, Y., Sato, M.: Humanscale haptic interaction with a reactive virtual human in a real-time physics simulator. Comput. Entertain. 4(3), 9 (2006)

[5] Heck, R., Gleicher, M.: Parametric motion graphs. In: I3D 2007: Proceedings of the 2007 symposium on Interactive 3D graphics and games, pp. 129–136. ACM, New York (2007)

[6] Ikemoto, L., Arikan, O., Forsyth, D.: Learning to move autonomously in a hostile world. In: SIGGRAPH 2005: ACM SIGGRAPH 2005 Sketches, p. 46. ACM, New York (2005)

[7] Ikemoto, L., Arikan, O., Forsyth, D.: Knowing when to put your foot down. In: I3D 2006: Proceedings of the 2006 symposium on Interactive 3D graphics and games, pp. 49–53. ACM, New York (2006)

[8] Kovar, L., Gleicher, M.: Flexible automatic motion blending with registration curves. In: Proceedings of the 2003 ACM SIGGRAPH/ Eurographics Symposium on Computer Animation, pp. 214–224, Eurographics Association (2003)

[9] Kovar, L., Gleicher, M., Pighin, F.: Motion graphs. ACM Trans. Graph. 21(3), 473–482 (2002)

[10] Kovar, L., Schreiner, J., Gleicher, M.: Footskate cleanup for motion capture editing. In: Proceedings of the ACM SIGGRAPH symposium on Computer animation, pp. 97–104. ACM Press, New York (2002)

[11] Lai, Y.-C., Chenney, S., Fan, S.-H.: Group motion graphs. In: SCA 2005: Proceedings of the 2005 ACM SIGGRAPH/Eurographics symposium on Computer animation, pp. 281–290. ACM, New York (2005)

[12] Lee, K.H., Choi, M.G., Lee, J.: Motion patches: buildings blocks for virtual environments annotated with motion data. In: SIGGRAPH 2005: ACM SIGGRAPH 2005 Sketches, p. 48. ACM, New York (2005)

[13] Liu, C.K., Popovic´, Z.: Synthesis of complex dynamic character motion from simple animations. In: Proceedings of the 29th annual conference on Computer graphics and interactive techniques, pp. 408–416. ACM Press, New York (2002)

[14] Park Il, S., Shin, H.J., Shin, S.Y.: On-line locomotion generation based on motion blending. In: Proceedings of the 2002 ACM SIGGRAPH/Eurographics symposium on Computer animation, pp. 105–111. ACM Press, New York (2002)

[15] Pullen, K., Bregler, C.: Motion capture assisted animation: texturing and synthesis. In: Proceedings of the 29th annual conference on Computer graphics and interactive techniques, pp. 501–508. ACM Press, New York (2002)

[16] Ren, L., Patrick, A., Efros, A.A., Hodgins, J.K., Rehg, J.M.: A data-driven approach to quantifying natural human motion. ACM Trans. Graph. 24(3), 1090–1097 (2005)

[17] Sung, M., Kovar, L., Gleicher, M.: Fast and accurate goal-directed motion synthesis for crowds. In: SCA 2005: Proceedings of the 2005 ACM SIGGRAPH/Eurographics symposium on Computer animation, pp. 291–300. ACM, New York (2005)

[18] Witkin, A., Popović, Z.: Motion warping. In: 29th Annual Conference Series on Computer Graphics, pp. 105–108 (1995)

[19] Tak, S., Song, O.y., Ko, H.-S.: Spacetime sweeping: an interactive dynamic constraints solver. In: Proceedings of Computer Animation, 2002 (2002)

# 4

# 3D Object Retrieval: Inter-Class vs. Intra-Class*

Theoharis Theoharis

Department of Informatics, University of Athens,
Panepistimiopolis, 15784,
Athens, Greece
theothoe@di.uoa.gr
http://graphics.di.uoa.gr

**Summary.** The monotonic advances in the accuracy of 3D object retrieval methods has created a euphoria and the belief that 100% accuracy is a matter of time. However, accuracy is tied to datasets. Algorithmic progress in this field is often linked to the availability of datasets. A new, harder, dataset will often derail existing algorithms and create a new research wave. We attempt to define and compare inter- and intra-class 3D object retrieval and show that there is a huge performance gap between them. We present relevant bibliography and our own inter- and intra-class retrieval methods and applications.

**Keywords:** 3D Object Retrieval, inter-class, intra-class, shape descriptor, biometrics.

## 4.1 Introduction and Motivation

The increasing availability of 3D objects makes content based retrieval a key operation. Retrieval methods are based on the creation of a *shape descriptor* that uniquely describes the shape of the objects in an efficient manner. Ideally a shape descriptor should be invariant to similarity transformations (scale, translation and rotation). In addition, it should be compact in order to reduce storage requirements and support rapid computation of the similarity between objects.

For the purposes of this paper we define a *class* of objects as a set of objects with similar shape. From this definition it follows that we can construct a *class model* that captures the basic shape characteristics of the class. Examples of classes are the class of human faces and the class of passenger cars.

We can distinguish 3D object retrieval (OR) methods as *intra-class* and *inter-class* according to whether they operate on a single class of objects or across classes (see Figure 4.1). An intra-class retrieval method needs far greater discriminative power than an inter-class method because it has to decide between very similar objects. An intra-class method can take advantage of the class model in order to boost its accuracy. For example an intra-class method for human face retrieval (which addresses the well known problem of *face recognition*) can fit a

---

* This invited talk is based on joint work with Prof. I. A. Kakadiaris, Drs I. Pratikakis and S. Perantonis, G. Passalis, P. Papadakis and G. Toderici.

D. Plemenos, G. Miaoulis (Eds.): Arti. Intel. Techn. for Comp. Graph., SCI 159, pp. 55–66.
springerlink.com

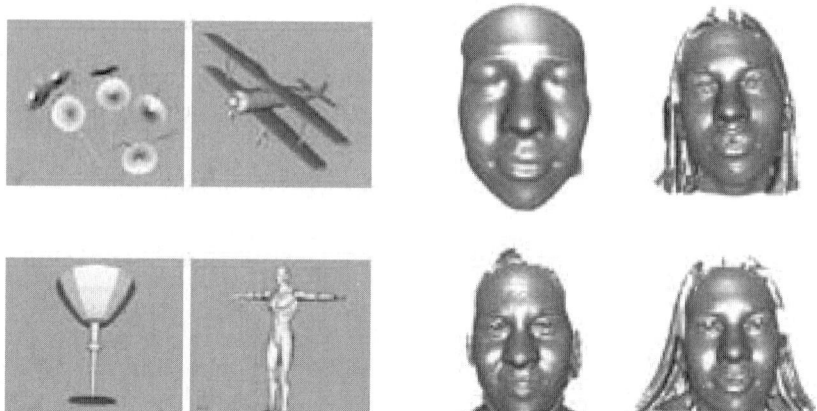

**Fig. 4.1.** Inter-Class (left) vs Intra-Class Datasets (right)

3D facial model onto a subject's 3D facial scan and thus determine the location of useful features on the subject's face.

The motivation behind this talk is to mark and quantify the distinction between inter- and intra-class OR methods. To this end we present a state-of-art (SOA) inter-class method and, after giving its (high) performance on inter-class datasets, we show how poorly it performs on an intra-class dataset. We then present a SOA intra-class method which performs extremely well on the same intra-class dataset, but of course is not applicable at all to inter-class problems.

The inter-class and intra-class OR methods used for this comparison are based on previous work as described in [17, 12, 13] (inter-class) and [16, 10] (intra-class); these are representative samples of retrieval work which all achieved SOA performance.

## 4.2  Inter-Class OR Methods

Inter-class OR methods may be classified into two broad categories according to their dimensionality, 2D and 3D. A central issue in all OR methods is *invariance under rigid transformations* (rotation and translation), which ensures that an object will be correctly retrieved regardless of pose. To this effect, several *pose normalization* techniques have been used. Commonly, a standard rigid pose is applied to an object before computing its shape descriptor.

*2D methods* use two-dimensional images/projections of 3D objects which may be renderings from different viewpoints, depth buffers or contours. Similarity is then measured using 2D matching techniques. Chen et al. [5] proposed the Light Field descriptor, which is comprised of Zernike moments and Fourier coefficients computed on a set of projections taken from the vertices of a dodecahedron. Vranic [22] proposed a shape descriptor where features are extracted from depth buffers produced by six projections of the model, one for each side of a cube

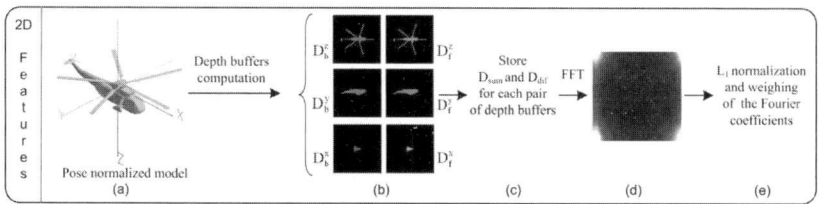

**Fig. 4.2.** 2D Feature Extraction in [17]

which encloses the model. In the same work, the Silhouette-based descriptor is proposed which uses the silhouettes produced by the three projections taken from the Cartesian planes. Zarpalas et al. [24] introduce the Spherical Trace Transform descriptor, which is an extension to 3D of the 2D Trace Transform.

In previous work [17] we proposed PTK, a depth buffer based descriptor which uses parallel projections to capture the models shape. In Figure 4.2 the main steps (a)-(e) for the extraction of this 2D shape descriptor are depicted. Pose normalization (see [13]) is achieved by:

- Computing the centroid of a model and translating it to the coordinate origin (translation normalization) and
- using Principal Component Analysis (PCA) on both the surface (CPCA) and the normals (NPCA) of the object (rotation normalization). We thus take into account both the surface area distribution and the surface orientation distribution that both characterize the shape of an object.

Two pose normalized versions of an object thus result, one for CPCA and one for NPCA, and they are processed in parallel. It has been shown that the combination of the two pose normalization strategies gives higher retrieval accuracy on most datasets [13]. From the pose normalized 3D model, we acquire a set of six depth buffers on the GPU by projecting the model to the faces of a cube which is centered at the origin and whose size is proportional to the models scale. For each pair of parallel depth buffers, we compute the difference $D_{diff} = D_f - D_b$ between the front ($D_f$) and back ($D_b$) depth buffer to capture the models thickness. To avoid information loss, the sum $D_{sum} = D_f + D_b$ of the front and back depth buffer is also computed. We next compute the Discrete Fourier Transform of $D_{diff}$ and $D_{sum}$ and normalize the coefficients to their unit $L_1$ norm. Two weighting schemes are finally applied to the coefficients; the first weighs them inversely proportionally to their degree, since lower frequencies are considered to have more information compared to higher frequencies which are more sensitive to noise; the second weighs them proportionally to the ranks of the principal directions. The final 2D shape descriptor of a model $i$ is the concatenation of the weighted coefficients of the three directions, for each rotation normalization method i.e. $2dsd_i^j$ where $j \in \{CPCA, NPCA\}$.

*3D methods* extract shape descriptors from the 3D shape-geometry of the 3D model and the similarity is measured using appropriate representations in the

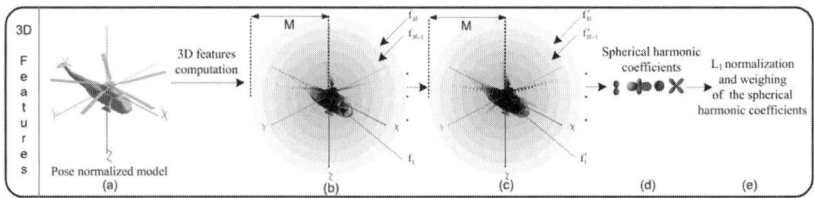

**Fig. 4.3.** 3D Feature Extraction in [12]

spatial domain or in the spectral domain. Ankerst et al. [1] proposed the Shape Histograms descriptor where 3D space can be divided into concentric shells, sectors, or both and for each part the models shape distribution is computed, giving a sum of histograms bins. Vranic [22] proposed the Ray-based descriptor which characterizes a 3D model by a spherical extent function capturing the furthest intersection points of the models surface with rays emanating from the origin. Spherical harmonics or moments can be used to represent the spherical extent function. A generalization of the previous approach is also described in [22], that uses several spherical extent functions of different radii. The GEDT descriptor proposed by Kazhdan et al. [9] is a volumetric representation of the Gaussian Euclidean Distance Transform of the 3D model's volume, expressed by norms of spherical harmonic frequencies.

In previous work [12] we proposed the CRSP shape descriptor. CRSP extracts 3D features using a spherical function-based representation of the 3D model and computes the spherical harmonics transform for each of the N spherical functions. Figure 4.3 depicts the main steps (a)-(e) for the extraction of this 3D shape descriptor. Pose normalization is achieved exactly as for PTK, resulting in two pose normalized versions (for CPCA and NPCA) which are processed in parallel. The surface of a 3D model is represented using a set of spherical functions, by projecting approximately equidistant parts of the 3D model to concentric spheres of increasing radii. This is done by computing the intersections of the models surface with rays emanating from the origin at directions $(m, n)$, where $m$ corresponds to the longitude and $n$ to the latitude coordinate. We next modify the spherical functions to include every point that is closer to the origin than the furthest intersection point on the respective ray. Thus, we obtain a volumetric, star-shaped representation of the 3D model and for each spherical function we compute the spherical harmonics transform, keep a subset of the spherical harmonic coefficients and normalize them to their unit $L_1$ norm. Finally, the coefficients are weighed inversely proportionally to their degree. The final 3D shape descriptor of a model $i$ is the concatenation of the weighted coefficients of the $N$ spherical functions, for each rotation normalization method i.e. $3dsd_i^j$ where $j \in \{CPCA, NPCA\}$.

*Hybrid* descriptors which combine 2D and 3D characteristics have also been proposed. The key issue here, as when combining any descriptors, is to ensure

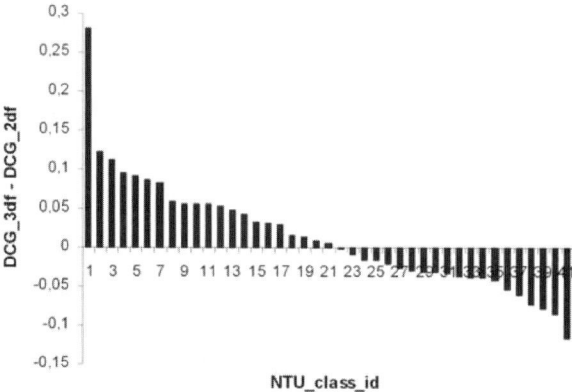

**Fig. 4.4.** Complementarity of 2D PTK and 3D CRSP descriptors, from [13]

that the descriptors being combined have *complementary* characteristics. Ideally they should have disjoint failure sets.

We have investigated the complementarity of the 2D PTK and the 3D CRSP shape descriptors on the NTU dataset (National Taiwan University Benchmark, 549 models, http://3D.csie.ntu.edu.tw/), see Figure 4.4. The vertical axis shows the difference between the Discounted Cumulative Gain (DCG) score of the two descriptors on the NTU dataset. The DCG score is a commonly used quantitative performance measure [20] which is computed by weighing correct results according to their position in the retrieval list (the maximum score is 100%). The horizontal axis is an id-number corresponding to a specific class of 3D model in the NTU dataset. Given that the positives and negatives in the Figure are roughly equal, we can conclude that the chosen 2D and 3D shape descriptors have a complementary behavior, that is, there are classes where the 3D features are more discriminative than the 2D features and vice versa. We have thus proposed a hybrid descriptor based on PTK and CRSP [13].

Hybrid descriptors generally exhibit the best OR performance. Having chosen the ingredients of a hybrid descriptor, it is then crucial to use an effective *fusion technique* to combine their results. Such techniques can be simple min-max or normalize and sum rules, but more advanced methods can give better results, see [19]. We have opted for a simple sum strategy of the 2D and 3D shape descriptors over the minimum of the two pose normalization strategies. In other words, the distance between two models $p$ and $q$ is computed as:

$$dist(p,q) = \min_{j\in\{CPCA,NPCA\}}(L_1(2dsd_p^j, 2dsd_q^j)) + \min_{j\in\{CPCA,NPCA\}}(L_1(3dsd_p^j, 3dsd_q^j)).$$

$L_1$ is the Manhattan distance and 2D and 3D coefficients are pre-normalized to [0,1]. A search engine based on our descriptor is available on-line at http://emedi3.emedi.iit.demokritos.gr/cil3d/.

### 4.2.1 Performance of Inter-Class OR Methods on Inter-Class Datasets

Two key aspects when making retrieval-accuracy comparisons between rival shape descriptors are:

- The dataset used and
- The test used.

It is important to use widely accepted and freely available datasets in order to make robust and repeatable judgements. Common tests include Precision-Recall diagrams and Cumulative Match Characteristic (CMC) curves. The former diagrams plot the percentage of erroneously retrieved objects against the percentage coverage of the class that should have been retrieved. CMC curves plot the percentage of objects that are correctly retrieved within the $i^{ith}$ rank. Apart from the NTU dataset described above (549 models), three other well known, freely available and relatively large datasets are the Princeton Shape Benchmark (PSB) (907 models), the CCCC dataset from University of Konstanz (472 models) and the MPEG-7 dataset (1300 models). Figure 4.5 compares the Hybrid shape descriptor just described against 3 state-of-art descriptors, the Light Field (LF) descriptor [Chen03], the spherical harmonic representation of the Gaussian Euclidean Distance Transform (SH-GEDT) descriptor [Kazh03] and the DSR472 descriptor [Vran04a] using precision-recall diagrams. As can be clearly seen the Hybrid descriptor outperforms all previous methods on these datasets.

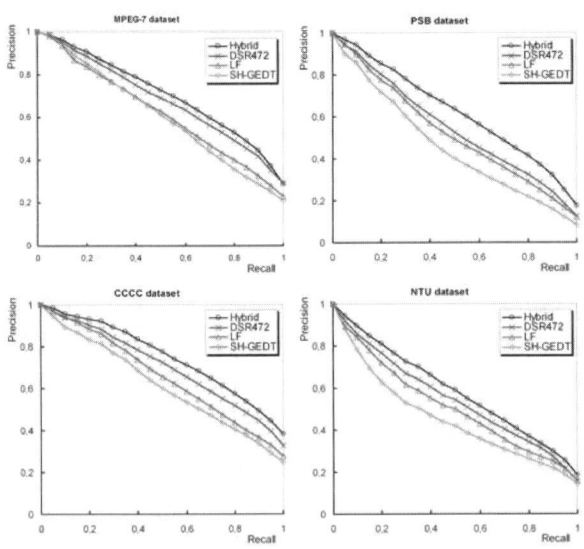

**Fig. 4.5.** Precision-Recall Diagrams for Hybrid Shape Descriptor and other State-of-Art Descriptors, from [12]

## 4.3   Inter-Class OR Methods on Intra-Class Dataset

An observation from Figure 4.5, is that inter-class methods have achieved high retrieval accuracy and a state of maturity. This can be deduced from the fact that there are no sudden leaps in performance and new methods only push the accuracy curve a little higher. It is therefore particularly tantalizing to check out how such methods would fare on an intra-class dataset, where the objects are very similar to each other. Could they do as well and thus solve many open problems, such as 3D Face Recognition?

### 4.3.1   Performance of Inter-Class OR Methods on Intra-Class Dataset

To test the above we employed FRGC v2, the largest publicly available 3D facial dataset which was created for the latest Face Recognition Grand Challenge organized by US National Institute for Standards and Technology. FRGC v2 consists of 4007 range images from 466 subjects; many have facial expressions and pose variations. The range images are first converted to a polygonal representation. Figure 4.6 is a precision-recall diagram comparing several SOA inter-class 3D OR methods on FRGC v2. Specifically, DSR, DBD, SIL, and RSH refer to hybrid-based, depth buffer-based, silhouette-based, and ray-based feature vectors provided by D. Vranic [22, 23]. Our hybrid method is not in the comparison as it was not available at the time of comparison, but no doubt it would achieve similar (hopefully better!) performance. This is in stark contrast to the performance of an intra-class method (FWV) on the same dataset. The performance gap is worth a thousand words. FWV is the intra-class method which will be described in the following section. This comparison does not discredit inter-class

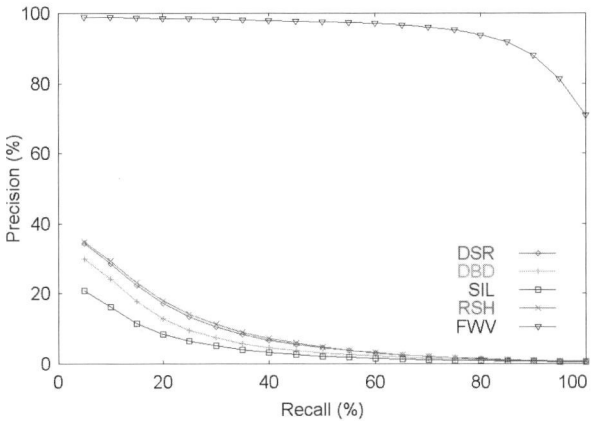

**Fig. 4.6.** Inter-Class OR Methods on Intra-Class Dataset (DSR,DBD,SIL,RSH) and an Intra-Class Method on the same Dataset (FWV), from [16]

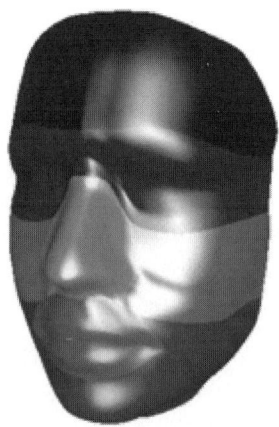

**Fig. 4.7.** Annotated Face Model for Intra-Class OR, from [16]

methods which offer generality and can be applied to virtually any 3D OR task; intra-class methods sacrifice this generality by concentrating on a single class in order to discriminate among minute differences of intra-class datasets, such as human faces.

## 4.4  Intra-Class OR Methods

Let us define a single-class dataset as one for which a class model can be built that captures the basic characteristics of the class. The goal in intra-class OR is to find the subtle differences among objects that belong to a single class, and thus share many common features. Even though intra- and inter-class OR belong to the same domain, the methods that tackle each of them have a different focus. Intra-class methods can make use of the class model in order to boost their accuracy, while at the same time lose their generality.

Much of the recent work in intra-class OR has arisen from the field of biometrics, mainly due to recent research funding; good surveys are [2, 3, 4, 25]. However, such methods are generally limited to a single biometric class (e.g. human ears).

We have developed a general intra-class OR method [16]. While it is applicable to biometric problems and has shown SOA performance in both face and ear recognition, it can also be applied to other object classes as well. Our method uses an Annotated Model (AM) of the class, which is constructed offline, and represents a typical object of the class. The annotation into different areas allows the separate treatment of these areas. The AM has a regularly sampled mapping from $R^3$ to $R^2$ (UV parameterization); this allows it to be converted to an equivalent 2D representation called a *geometry image* [6, 7]. Figure 4.7 shows our AM for the class of human faces. It is annotated into areas that can later be assigned different weights, based on their reliability. The AM can be constructed by taking the average of a number of class models.

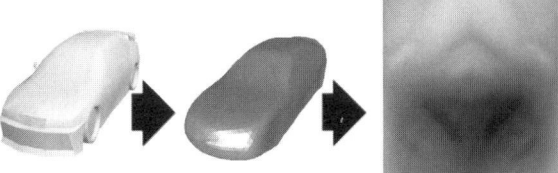

**Fig. 4.8.** Original Object → Fitted AM → Geometry Image, from [16]

The shape descriptor of an object O is extracted by taking the following steps:

- **Registration:** Register O with the AM to achieve pose normalization
- **Fitting:** Deform AM to fit O
- **Geometry Image:** Convert the deformed AM to a geometry image
- **Wavelet Analysis:** Apply wavelet analysis to the geometry image

A portion of the wavelet coefficients (which correspond to the lower frequencies) are then kept as the shape descriptor.

For the registration we have developed a 3-stage technique which is based on spin images, ICP and simulated annealing on depth buffers. Each stage uses as input the output of the previous one. Later stages are more sensitive to initial conditions but also provide more accurate registration. Registration is the most crucial step of the method as registration errors will propagate to subsequent stages and cause total failures.

The AM is next fitted to O in order to capture the geometric differences between the two. The AM is an elastically adapted deformable model based on the work of Metaxas and Kakadiaris [11] and is represented as a subdivision surface. As the AM is deformed to fit O, a detailed map of differences between the two is constructed [8, 14]. This map uniquely characterizes O.

The deformed AM is then converted to a geometry image, which is possible due to the construction of the AM. A 2D representation of the characteristic map of O is thus created. The geometry image contains one channel per coordinate component $(x, y, z)$ and can be stored as an RGB image, Figure 4.8. The geometry image is then converted to the wavelet domain in order to create a more robust representation of O. The Haar filter bank is used [21] due to its localization properties, which allow us to identify the wavelet coefficients that correspond to different areas of the AM. It was experimentally determined that a small subset of the wavelet coefficients convey most of the information present in the geometry image; these correspond to the lower frequencies.

Let $\alpha$ be the mask of stored wavelet coefficients of an area of the AM and $A$ be the set of all such masks for the entire AM. If $O_1$ and $O_2$ are the wavelet representations of two objects, the similarity score of the two objects in the area of $\alpha$ for channel $x$ of the geometry image is:

$$S_x(O_1, O_2, \alpha) = \sum_{w \in \alpha} |O_1[w] - O_2[w]|$$

The similarity score for all channels is then:

$$S(O_1, O_2, \alpha) = S_x(\alpha) + S_y(\alpha) + S_z(\alpha)$$

And the similarity score over all the areas of the two objects is:

$$S(O_1, O_2) = \sum_{\alpha \in A} w_\alpha \cdot S(O_1, O_2, \alpha)$$

where $w_\alpha$ is the weight that corresponds to the area of $\alpha$.

### 4.4.1    Performance of Intra-Class OR Methods on Intra-Class Datasets

We have applied our intra-class retrieval method to the problems of face and ear recognition, by creating the appropriate Annotated Models and experimentally determining weights for their areas. Figure 4.9 shows its performance on the FRGC v2 3D face dataset [18], which is the largest and most widely acknowledged one to date. It consists of 4007 3D facial scans from 466 subjects. A verification scenario is assumed here where the vertical axis plots the portion of correct verifications against various false accept rates. It is interesting to note that facial expressions do not significantly degrade performance, a quality that can be attributed to the use of the deformable model. The reported performance was top at the time of the initial publication and we believe that it still remains so.

Figure 4.10 shows the performance of our intra-class OR method on two 3D ear databases in a verification scenario where the vertical axis gives the identification score for various ranks of the horizontal axis. The larger the rank allowed the greater the possibility that the correct subject is identified. This performance was state-of-art in ear recognition at the time of publication; only one other method performed slightly better.

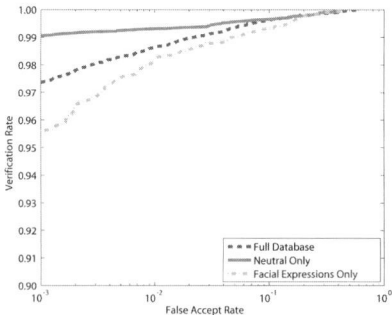

**Fig. 4.9.** Performance of Intra-Class Method on FRGC v2 Facial Dataset, from [10]. Receiver Operating Characteristics (ROC) curve. Note that the vertical axis starts from 0.90.

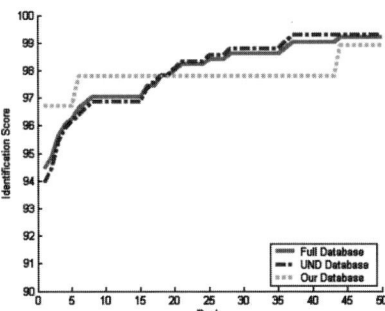

**Fig. 4.10.** Performance of Intra-Class Method on Databases of 3D Ear Scans, from [15]. The UND database consists of 830 Ear datasets from 415 subjects and our (UH) database consists of 201 Ear datasets from 110 subjects. Note that the vertical axis starts from 0.90.

The interesting thing however was the ease with which the generic intra-class method was applied to 3D face and then to 3D ear recognition. It simply involved constructing and tuning the respective Annotated Model of the class.

As the number of shape descriptors grows and more datasets become available, object retrieval contests are beginning to be organized. Some deal with intra-class problems such as biometrics (e.g. Face Recognition Grand Challenge, Face Recognition Vendor Test) while others take a more general approach and deal mainly with inter-class problems (e.g. SHREC).

# References

1. Ankerst, M., Kastenmuller, G., Kriegel, H.P., Seidl, T.: Nearest neighbor classification in 3D protein databases. In: ISMB, pp. 34–43 (1999)
2. Bowyer, K., Chang, K., Flynn, P.: A Survey of Approaches to 3D and Multi-Modal 3D-2D Face Recognition. In: Proc. IEEE Intl. Conf. Pattern Recognition, pp. 358–361 (August 2004)
3. Chang, K.I., Bowyer, K.W., Flynn, P.J.: An Evaluation of Multi-Modal 2D-3D Face Biometrics. IEEE Trans. Pattern Analysis and Machine Intelligence 27(4), 619–624 (2005)
4. Chang, K., Bowyer, K., Flynn, P.: A Survey of Approaches and Challenges in 3D and Multi-Modal 2D-3D Face Recognition. Computer Vision and Image Understanding 101(1), 1–15 (2006)
5. Chen, D.Y., Tian, X.P., Shen, Y.T., Ouhyoung, M.: On visual similarity based 3D model retrieval. Eurographics. Computer Graphics Forum, 223–232 (2003)
6. Gu, X., Gortler, S., Hoppe, H.: Geometry Images. In: Proc. SIGGRAPH, pp. 355–361 (July 2002)
7. Kakadiaris, I.A., Shen, L., Papadakis, M., Kouri, D., Hoffman, D.: g-HDAF Multiresolution Deformable Models for Shape Modeling and Reconstruction. In: Proc. British Machine Vision Conf., pp. 303–312 (September 2002)

8. Kakadiaris, I., Passalis, G., Theoharis, T., Toderici, G., Konstantinidis, I., Murtuza, N.: Multimodal Face Recognition: Combination of Geometry with Physiological Information. In: Proc. IEEE Computer Vision and Pattern Recognition, pp. 1022–1029 (June 2005)
9. Kazhdan, M., Funkhouser, T., Rusinkiewicz, S.: Rotation invariant spherical harmonic representation of 3D shape descriptors. In: Eurographics/ACM SIGGRAPH symposium on Geometry processing, pp. 156–164 (2003)
10. Kakadiaris, I.A., Passalis, G., Toderici, G., Murtuza, N., Lu, Y., Karampatziakis, N., Theoharis, T.: 3D face recognition in the presence of facial expressions: An annotated deformable model approach. IEEE Transactions on Pattern Analysis and Machine Intelligence (TPAMI) 29(4), 640–649 (2007)
11. Metaxas, D., Kakadiaris, I.: Elastically Adaptive Deformable Models. IEEE Trans. Pattern Analysis and Machine Intelligence 24(10), 1310–1321 (2002)
12. Papadakis, P., Pratikakis, I., Perantonis, S., Theoharis, T.: Efficient 3D Shape Matching and Retrieval using a Concrete Radialized Spherical Projection Representation. Pattern Recognition 40(9), 2437–2452 (2007)
13. Papadakis, P., Pratikakis, I., Theoharis, T., Passalis, G., Perantonis, S.: 3D Object Retrieval using an Efficient and Compact Hybrid Shape Descriptor. In: Eurographics Workshop on 3D Object Retrieval, pp. 9–16 (2008)
14. Passalis, G., Kakadiaris, I., Theoharis, T., Toderici, G., Murtuza, N.: Evaluation of 3D Face Recognition in the Presence of Facial Expressions: An Annotated Deformable Model Approach. In: Proc. IEEE Workshop Face Recognition Grand Challenge Experiments, pp. 171–179 (June 2005)
15. Passalis, G., Kakadiaris, I.A., Theoharis, T., Toderici, G., Papaioannou, T.: Towards fast 3D ear recognition for real-life biometric applications. In: Proc. of IEEE International Conference on Advanced Video and Signal based Surveillance (AVSS 2007), London, UK (September 2007)
16. Passalis, G., Kakadiaris, I.A., Theoharis, T.: Intra-class retrieval of non-rigid 3D objects: Application to Face Recognition. IEEE Transactions on Pattern Analysis and Machine Intelligence (TPAMI) 29(2), 218–229 (2007)
17. Passalis, G., Theoharis, T., Kakadiaris, I.A.: PTK: A Novel Depth Buffer-Based Shape Descriptor for Three-Dimensional Object Retrieval. The Visual Computer 23(1), 5–14 (2007)
18. Phillips, P., Flynn, P., Scruggs, T., Bowyer, K., Chang, J., Hoffman, K., Marques, J., Min, J., Worek, W.: Overview of the Face Recognition Grand Challenge. In: Proc. IEEE Computer Vision and Pattern Recognition, pp. 947–954 (June 2005)
19. Ross, A., Nandakumar, K., Jain, A.K.: Score Level Fusion. In: Handbook of Multibiometrics, ch. 4. Springer, Heidelberg (2006)
20. Shilane, P., Min, P., Kazhdan, M., Funkhauser, T.: The princeton shape benchmark. In: Shape Modeling International, pp. 167–178 (2004)
21. Stollnitz, E., DeRose, T., Salesin, D.: Wavelets for Computer Graphics: Theory and Applications. Morgan Kaufmann, San Francisco (1996)
22. Vranic, D.: 3D Model Retrieval, PhD dissertation, Universitat Leipzig (May 2004)
23. Vranic, D.: Content-Based Classification of 3D-Models by Capturing Spatial Characteristics (2004), http://merkur01.inf.unikonstanz.de/CCCC/
24. Zarpalas, D., Daras, P., Axenopoulos, A., Tzovaras, D., Strintzis, M.G.: 3D model search and retrieval using the spherical trace transform. EURASIP Journal on Advances in Signal Processing (2007)
25. Zhao, W., Chellappa, R., Phillips, P.J., Rosenfeld, A.: Face Recognition: A Literature Survey. ACM Computer Surveys 35(4), 399–458 (2003)

# 5

# Improving Light Position in a Growth Chamber through the Use of a Genetic Algorithm

Samuel Delepoulle[1], Christophe Renaud[1], and Michaël Chelle[2]

[1] Laboratoire d'Informatique du Littoral (LIL), BP 719, 62228
   Calais, France
   `{delepoulle,renaud}@lil.univ-littoral.fr`
[2] INRA, UMR 1091 Environnement et Grandes Cultures, 78850
   Thiverval-Grignon, France
   `chelle@grignon.inra.fr`

**Summary.** Growth chambers are used by agronomists for various experiments on plants. Because light impinging on plants is one of the main parameters of their growth, inhomogeneity in light reception can provide large bias in the experiments. In this paper we present the first steps of a framework which aims at computing the best locations of light sources that could ensure an homogeneous lighting at some places in these chambers. For this purpose we extent the capabilities of a global illumination approach dedicated to growth chambers. We use Genetic Algorithms with simple source and material models in order to find good light sources locations. Our first results show that it is possible to find such interesting location and that they improve the lighting distribution on the experiment tables used in these chambers.

## 5.1 Introduction

From the point of view of computer graphics users photo-realistic images can be generated from algorithms solving the rendering equation [15]. Amongst the lot of such algorithms that have been developed during the last two decades, some have been proved to be both efficient and accurate for taking into account general light sources and materials properties [17, 18, 22, 23]. Despite their high computational requirements, they even become commonly in use in the computer graphics users community. Because these algorithms are physically based they do not only compute pixels color. They have the great advantage of being able to physically compute the quantity of light received on and reflected by any surface. Thus they can be used not only for image generation purpose but for lighting simulation too.

Amongst the various scientific domains that can be interested in such simulations, agronomy is concerned with the estimate of light impinging on plants. Indeed the leafs and stems irradiance has a great effect on biomass production through the photosynthesis process and on the way plants grow through

D. Plemenos, G. Miaoulis (Eds.): Arti. Intel. Techn. for Comp. Graph., SCI 159, pp. 67–82.
springerlink.com                                    © Springer-Verlag Berlin Heidelberg 2009

photo-morphogenesis. This estimate is even critical when agronomists compare
different species inside growth chamber. Indeed reflections on the chamber walls
or localisation inside the room can change dramatically the irradiance of plant
organs. Thus results are often biased and hardly reproducible in different chambers, and even in the same chamber using a different plant positioning. Some
models have been previously proposed [1, 2] but they mainly rely on the tur-
bid medium approach, which does not enable the estimate of the irradiance of
individual plants or organs. We thus have developed a simulation framework,
$SEC_2$[1], which is derived from a global illumination algorithm and allows users
to accurately compute irradiance estimate on 3D plants. This framework has
been shown to be accurate enough to simulate some existing growth chambers
even with some simplifications.

In this paper we are interested in extending the capabilities of $SEC_2$ for
a reverse problem [14]. We would like our framework to be able to provide
some information on light sources choice and location according to some given
constraints, for example ensuring a homogeneous lighting in some place of the
chamber. This kind of constraint is of main importance for agronomists because it
will reduce the variability of plants irradiance and the corresponding experiment
bias. We thus propose to research for optimal lights location through the use of
Genetic Algorithms (GAs) [11] which have proved their efficiency for solving
optimization problems [9]. In the next section we describe more accurately the
problems of growth chambers and summarize $SEC_2$ capabilities. We present in
section 5.3 the principles of GAs and we detail in section 5.4 our model for
optimal location research. Some results are presented and discussed in section
5.5 before some perspectives to this work are given.

## 5.2   Simulating Lighting in Growth Chamber

### 5.2.1   Growth Chambers

Agronomists use so called growth chambers for studying plant response to ge-
nomic and environmental changes. Such studies are required to develop function-
structure plant model, which enables the understanding of the biological complexity. For such studies, a well-controlled and spatially homogeneous climate is
necessary in order to be able to strictly compare results from different treatments
or genotypes. Growth chambers are thus designed for allowing users to control
temperature, lighting and hydrometry. However a large variety of growth chambers exists depending on their geometry, materials, lighting and control system
(see figure 5.1).

As a consequence each growth chamber should be considered as a unique
radiative system with its own lighting properties. Even if most of them have been
designed to minimize the spatial heterogeneity of lighting and more generally of
climate, measurements have shown that plants may experiment different lighting

---

[1] *"Simulation d'Eclairage en Chambre de Culture"* which means Growth Chamber
  Lighting Simulation.

**Fig. 5.1.** Four samples of existing growth chambers which highlight their variety

conditions depending on their location within the same chamber [20, 3]. Bias can then be introduced in the results of any experience.

### 5.2.2  Lighting Simulation

Because lighting measurements are often long and difficult to perform we developed a framework for growth chamber lighting simulation so called $SEC_2$. It allows agronomists to perform measures with various virtual captors and to situate those captors at any location in the virtual chamber. Based on the photon-mapping approach developed by Jensen [12, 13], photons are thrown from the light sources to the scene. They are reflected by the surfaces according to their BRDF until their absorption. Since none image has to be rebuilt photons impacts are recorded only on dedicated captors. Several kinds of captors are available and even plants (or parts of plants) can be used for recording incident photons (see figure 5.2). The number of photons that have reached any captor is then converted into graphical values in order to be able to visualize and to analyze the results. Figure 5.3 shows some outputs of the $SEC_2$ simulator.

Up to date $SEC_2$ uses some simplifications for computational power reasons. More specifically sources and/or complex lighting devices are approximated by

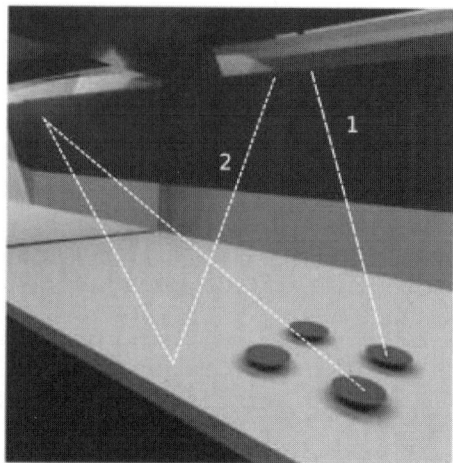

**Fig. 5.2.** $SEC_2$ captors record not only direct lighting (1) but also indirect one resulting of reflections (2)

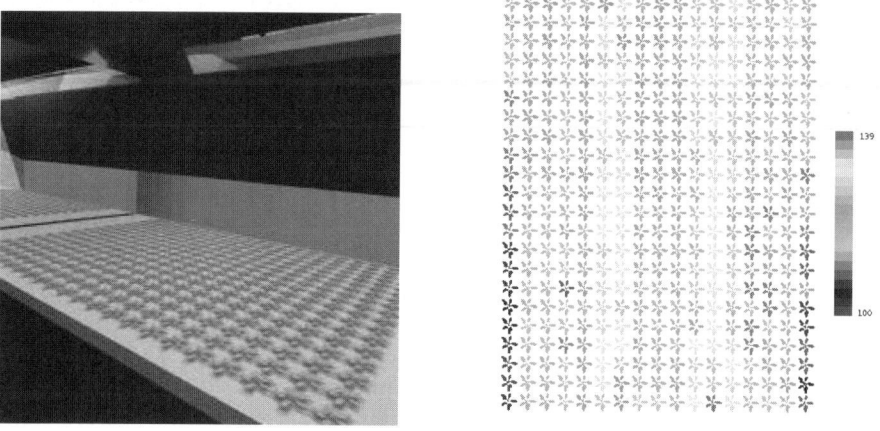

**Fig. 5.3.** Some examples of $SEC_2$ outputs : the geometry of the scene (left) ; a result of the average plant irradiance (right)

isotropic point or purely diffuse simple area light sources. BRDFs[2] are approximated by a Phong's reflection model the validity of which should be investigated for plants material. Nevertheless the ability to reproduce a virtual growth chamber as well as the correctness of $SEC_2$ were favorably assessed by comparing measured and simulated transversal profiles of irradiance at different elevations in an empty room [4].

---

[2] Bidirectional Reflectance Distribution Functions.

### 5.2.3   Reverse Lighting Problem

Once simulation highlights the problem of lighting heterogeneity in a given room it could be useful to provide tools that can help agronomists and/or growth chamber designers to correct this one. Due to the geometry of such experimentation rooms several parameters have to be taken into account :

- the number, the type and the location of light sources since they are the source of any illumination into this closed environment ;
- the geometry and the BRDF of the walls of the chamber : because rooms are closed and often of small size, reflection of light onto the wall can be important. Furthermore this indirect lighting is less directional than direct source one ;
- height of the experimentation table can vary for the largest chambers anyway ;
- the plant position on the experimentation table.

In this paper we are interested in solving the sub-problem of finding the location of light sources that can ensure a homogeneous irradiance on the experimentation table. We assume that all other parameters are known and stable. For this purpose we use the capabilities of $SEC_2$ to provide good estimates of such irradiance as the entry of an optimization method using genetic algorithms.

## 5.3   Evolutionary Algorithms

### 5.3.1   General Principles

Evolutionary algorithms are computation methods based on natural evolution. Mechanisms inspired by biological evolution of species are simulated and produce a research in a solution space. A solution can be considered as an individual and biological mechanisms such as breeding, mutations and selection are used to produce new solutions.

Several kinds of evolutionary algorithms can be described. The most famous form is Genetic Algorithm proposed by J.H. Holland [11]. We can also cite for example Genetic Programming where individuals are computer programs [16] or Learning Classifier Systems (LCS) where individuals are condition-action rules.

### 5.3.2   Genetic Algorithms

Genetic algorithms use a genetic code for representing a solution. This code uses a finite number of symbols (like DNA's four letters). The most frequent representation of a solution is an array of bits.

Algorithm 5.1 shows the standard GA mechanism. First a population is initialized often randomly. Every individual of the population is assessed using an objective or *fitness* function ($\phi$) which is a measure of the quality of a solution. Then genetic operators are used to produce the next generation :

- a selection operator which selects best individuals and takes part to the reproduction. A wheel selection mechanism is generally used: the probability of being selected is proportional to the value of the fitness function;
- a mutation (or variation) operator which modifies randomly a short fragment of the genetic code;
- a cross-over (or recombination) operator: two genetic codes are combined in order to produce a new solution. Mechanisms of one, two or multiple cross-over can be used.

---

**Algorithm 5.1.** General algorithm for standard GA

---

initialisation of population $P_0$.
**repeat**
    **Evaluation** of each individual of population $P_t$ by computing $\phi$.
    **Selection** of best individuals.
    Product new generation $P_{t+1}$ using **Mutation** and **Crossover**
**until** stop condition

---

### 5.3.3   GA for Optimization

GAs are used successfully in Artificial Intelligence for solving optimization problems [10]. They provide efficient and robust solutions particularly if the solution landscape is difficult to predict or noisy. As compared with local search, they are known to be less sensitive to local optima [6, 21]. Due to recombination, the research in the solution space is more efficient. GAs are known to outperform conventional optimization techniques when problems are difficult, non linear or noisy. They can be used for solving design problems [8]. That is the reason why we suggest to use GAs for solving the light position problem.

## 5.4   Light Placement with GA

### 5.4.1   Previous Work

Ferentinos and Albright [7] previously used GAs for optimal design of lighting systems. They used a genetic algorithm to select a pattern of light in a greenhouse. A complex fitness function is used to take into account not only the light uniformity but also the number of lights used and the cost of the power consumption. For simplifying the problem, they did three assumptions:

- light sources are considered as points;
- sources can only be placed on the inner vertices of a grid, providing a discretization of the space;
- only direct lighting is taken into account.

The last assumption is acceptable for greenhouse but not for growth chambers. Indeed, walls have a high reflectance (diffuse white, mirror) in order not to waste light energy. This makes the indirect lighting particularly important in growth chamber [3].

### 5.4.2 Gene Coding for Light Position

A genome represents the position of lights in the scene. In order to reduce the research space, some parameters have been presently fixed. The vertical position of every light is constant (20 cm from the ceiling). The position of an area light is determined by two floating point values which represent the position $(x,y)$ of the center of the surface light source in the horizontal plane (see figure 5.4).

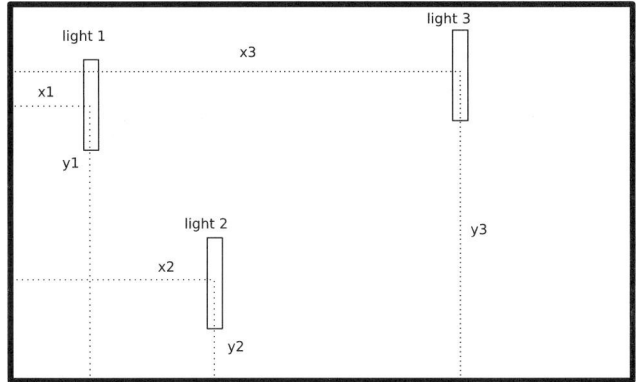

**Fig. 5.4.** The genome contains the code for lights position. Every light position can be represented by a pair of real numbers (x and y) relatively to the growth chamber coordinate system. The floating point information is then converted into a Boolean genetic code.

Every value is encoded onto 16-bits gene in order to apply genetics operators. A code of $b$ Boolean values can represent an integer number $m$ from 0 to $2^b - 1$. The coded value is then scaled in a determined interval (between $x_{min}$ and $x_{max}$) to compute the floating point value $(x)$.

$$x = x_{min} + \frac{M}{2^b - 1} \times (x_{max} - x_{min})$$

The precision $p$ of this coding is :

$$p = \frac{x_{max} - x_{min}}{2^b - 1}$$

For example the largest dimension of the real growth chamber we studied is

3.86 meters. With a 16 bits gene, the precision is lower than $6 \times 10^{-5}$ meters.

Classical form of mutation and cross-over is performed : with probability $p_{mut}$ a value is flipped; with probability $p_{cross}$ two genes are merged to produce a new code (see figure 5.5).

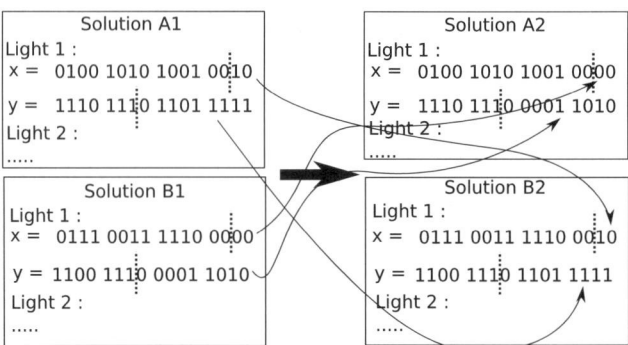

**Fig. 5.5.** Merging two solutions is performed by selecting a cross-over point and copying the genetic code of the parents. Two new light positions are created from two previous solutions.

### 5.4.3   Objective Function

In order to use GA for optimizing sources location, we have to decide which objective function is used. We research for the solution which produces the most homogeneous lighting on an experimentation plane. Uniformity can be viewed as the inverse of variance over the $n$ captors lying into the virtual growth chamber. We thus define the objective function $\phi$ as :

$$\phi = U = \frac{1}{\sigma^2}$$

with

$$\sigma^2 = \sum_{i=1}^{n} c_i^2 - \left( \frac{\sum_{i=1}^{n} c_i}{n} \right)^2$$

and $c_i$ the light intensity received by captor $i$.

In case of multiple lights we must consider the problem of overlapping sources. Light locations are represented by the position of their center which are initially chosen randomly. During the research process these locations are changed by the GA's operators. It is thus not possible to ensure that some light sources will not overlap. Because overlapping sources is not physically possible, we decide that the fitness value corresponding to this case is zero.

We used $SEC_2$ for computing the objective function corresponding to a solution (see figure 5.6) : for each source location, photons are thrown from the light sources and recorded by the captors. The objective value of each solution is computed and genetic algorithms are used to generate the new generation.

### 5.4.4   Testing Procedure

In order to see whether the algorithm is able to find interesting location for light sources, we used a very simplified model of growth chamber. It consists in an

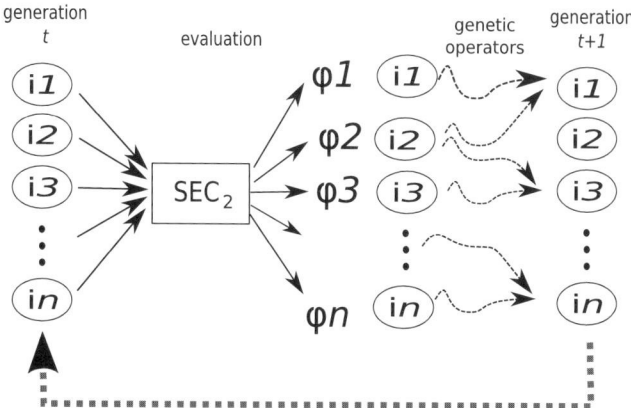

**Fig. 5.6.** $SEC_2$ engine is used to compute the quality of every solution in the current generation. Then genetic operators are used to produce the new generation.

empty room with walls, floor and ceiling. The material used for these surfaces is a grey color material and produces only diffuse reflections. In other words, the quantity of light reflected is the same in every direction. We used simple area sources (5×50 cm) and we performed runs with several number of sources ( 1, 2, 3, 4, 5, 6 or 10). Two growth chambers were tested with dimensions 4×4×4m or 6×4×4m.

Genetic Algorithm that has been used is the simple form with non-overlapping populations [9]. For each problem, 500 generations are computed. For each generation population included 20 individuals. The cross-over and mutation rates were set respectively to .6 and .01. The probability of cross-over must be relatively high for merging two solutions. The choice of mutation rate is difficult because variation allows exploration of solution space but tends to destroy good solutions.

Due to the stochastic nature of operators, there is no guarantee that the gene code of the best individual will be present in the next generation. This property is conform to biological evolution but can be a problem in design research. Thus we used the *elitism* property : the best individual is always stored and placed in the next generation. Using elitism thus ensures that a good solution cannot be lost. [5]

## 5.5 Results

### 5.5.1 Evolution of Solutions

Figures 5.7 and 5.8 show the result of 500 generation runs of 20 individuals respectively for the two previously described growth chambers. The source positions are initialized randomly at the beginning. We can see that GA's are able to

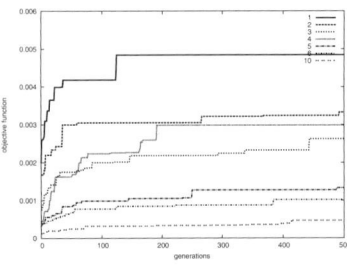
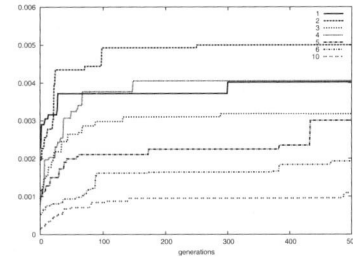

**Fig. 5.7.** Evolution over 500 generations. Growth chamber is 4×4×4 meters.

**Fig. 5.8.** Evolution over 500 generations. Growth chamber is 6×4×4 meters.

improve light position over generations in order to optimize uniformity of lighting. Curves named 1,2, 3, ..., 10 correspond to the number of searched light sources. Each one represents the best individual fitness ($\phi$).

The evolution of the fitness is more important at the beginning of the run. After 100 to 200 generations, a very good solution is generally found. The low evolution after tends to prove that a global optimum is found by the algorithm: the last part of the research is not efficient to discover effectively better solutions. Practical application may take into account this result by stopping the process more rapidly.

### 5.5.2   Comparison with Random Positioning

Although a random solution can not be considered as interesting it gives a kind of base level for knowing the gain produced by the algorithm.

Table 5.1 shows the fitness of GA's solution as compared to a random solution.

For each condition, average fitness over 20 individuals is computed. Globally the illumination's uniformity decreases with the number of sources. The GA

**Table 5.1.** Score of the best individual (GA solution) compared to the mean score of 20 random individuals (random solution)

| | 4×4×4 growth chamber | | | 6×4×4 growth chamber | | |
|---|---|---|---|---|---|---|
| sources | random solution | GA solution | gain | random solution | GA solution | gain |
| 1 | $2.02084\times10^{-3}$ | $4.8444\times10^{-3}$ | 2.397 | $1.9055\times10^{-3}$ | $4.0228\times10^{-3}$ | 2.111 |
| 2 | $.997806\times10^{-3}$ | $3.3234\times10^{-3}$ | 3.331 | $.98446\times10^{-3}$ | $4.9985\times10^{-3}$ | 5.077 |
| 3 | $.741305\times10^{-3}$ | $2.6273\times10^{-3}$ | 3.544 | $.78451\times10^{-3}$ | $3.1814\times10^{-3}$ | 4.055 |
| 4 | $.464485\times10^{-3}$ | $2.9837\times10^{-3}$ | 6.424 | $.43740\times10^{-3}$ | $4.0546\times10^{-3}$ | 9.270 |
| 5 | $.316875\times10^{-3}$ | $1.3316\times10^{-3}$ | 4.202 | $.31541\times10^{-3}$ | $3.0027\times10^{-3}$ | 9.520 |
| 6 | $.300934\times10^{-3}$ | $1.0219\times10^{-3}$ | 3.396 | $.29016\times10^{-3}$ | $1.9266\times10^{-3}$ | 6.640 |
| 10 | $.087835\times10^{-3}$ | $.47448\times10^{-3}$ | 5.402 | $.14083\times10^{-3}$ | $1.0940\times10^{-3}$ | 7.768 |

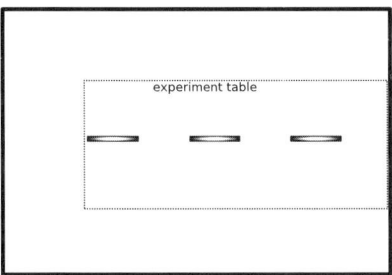

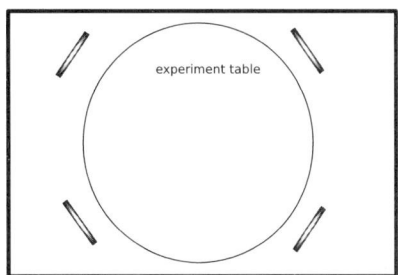

**Fig. 5.9.** Sources aligned along the central axis of a chamber

**Fig. 5.10.** Circular alignment of sources around the experiment table

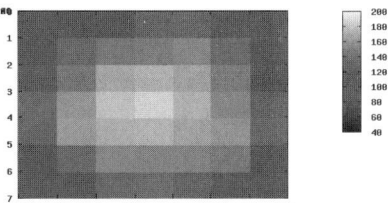

**Fig. 5.11.** Light distribution on the experiment table. Solution provided by GA after 500 generations. Three lights sources are positioned in a 6×4×4 growth chamber. $\phi = 3.169 \times 10^{-3}$

**Fig. 5.12.** Light distribution on the experimental table. Three lights are aligned (space between two lights is 1.5 meters) in the same growth chamber. $\phi = 0.911 \times 10^{-3}$

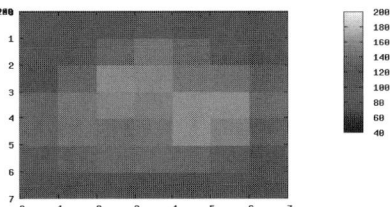

**Fig. 5.13.** Light distribution on the experimental table with three aligned lights. Two lights are spaced out 2 meters) in the same growth chamber. $\phi = 1.301 \times 10^{-3}$

provides a best solution in term of gain over random solution. In the case of improving the location of 4 or 5 light sources GA can even produce an enhancement factor of 10 : in other words, the variance over captors is divided by a 10 factor.

**Fig. 5.14.** View of the best individual after 500 generations for one, two or three area sources in cubical or rectangular growth chamber. Black disks are the captors which represent the location of the experimental table.

### 5.5.3    Comparison with Regular Distribution of Sources

The former results show that the GA is able to position sources with a best repartition of light than a random solution. But in the case of real growth chambers light are not set randomly. Rather they are generally located according to some regular schemes (see figures 5.9 and 5.10 for some examples).

We thus first compare the results of GA source locating with those of the linear source position of figure 5.10: figure 5.11 shows light distribution after adaptive selection of sources location for three sources. The result can be compared to light distributions when lights are aligned (figures 5.12 and 5.13). The GA's solution produces a more homogeneous distribution of light quantity.

Figure 5.14 shows the best solution found by GA for two growth chambers with 1, 2 or 3 area lights respectively. Note that for some chambers the location solution is not trivial even for a low number of sources.

### 5.5.4    Results in a More Complex Chamber

We then used a more complex model of growth chamber. The chamber used for this experiment is located at Grignon (France) and will be called the **Strader chamber** (see figure 5.15). Optical experiments were performed to estimate the BRDF of the chamber materials [3]. Basically the chamber is formed of :

- the room ($260\times390\times230$cm);
- an experiment table ($125\times300$cm) at a variable height;

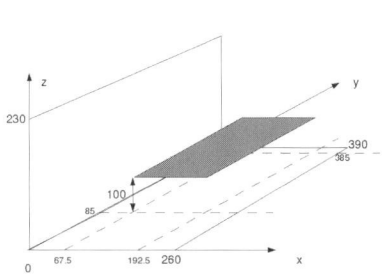 

**Fig. 5.15.** Two views of the strader growth chamber: the map of the chamber (left) and a view of the inside of the chamber with the experimentation table and the specular vertical reflector (right).

- a highly reflecting surface at the extremity of the table, used for simulating infinite plant canopies.

We used the same parameters as previously for the GA and the problem was to optimize the uniformity of lighting over the experiment table.

Figure 5.18 and 5.17 show the results of 500 generations of 20 individuals in the Strader's chamber. Homogeneity is clearly better with the GA positioning as compared to a linear distribution of light sources (figure 5.19 and 5.16): variance is divided by six. As a consequence, the algorithm tends to minimize the quantity of light received for the whole table : the values are aligned on the smallest.

Real lighting device of the strader growth chamber is composed by 18 cylindrical light bulbs. A highly specular reflector is placed over each light bulb. In the simulation the genetic algorithm selects the position of 18 area sources approximating the complex light sources (lamps and reflectors).

Table 5.2 shows the results for the present solution (two ramps of 9 ligths aligned over the experiment table) as compared to a random solution and to the genetic solution (500 generations of 20 individuals). The genetic method provides a more uniform lighting than the random solution. It's interesting to note that the current linear solution produce more heterogeneity than a random distribution.

The location of the light bulbs that have been provided by the GA are illustrated on figure 5.20 : it is clear that the proposed solution is far from being trivial.

**Table 5.2.** Fitness function computed in the strader chamber

| solution | random position | linear (current) | GA solution |
|---|---|---|---|
| fitness function | $1.658 \times 10^{-4}$ | $3.954 \times 10^{-5}$ | $2.732 \times 10^{-3}$ |

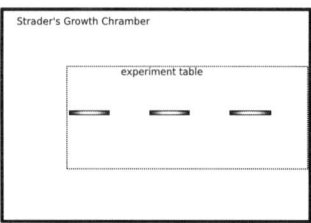

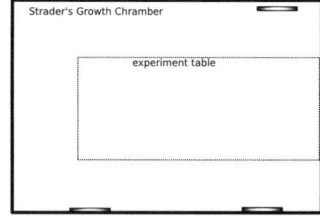

**Fig. 5.16.** Lights position in the Strader's growth chamber. The ligthts are aligned along the symetry axis of the chamber and over the experiment table.

**Fig. 5.17.** Lights position found by the Genetic Algorithm. Lights are placed near the wall increasing the indirect lighting on the experiment table.

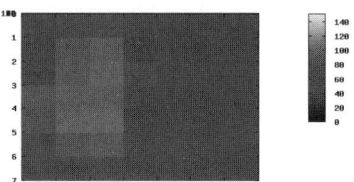

**Fig. 5.18.** Light distribution on the experimental table. Three lights are aligned (space between the center of two lights is 1 meter) in the Strader's growth chamber. $\phi = 2.27 \times 10^{-3}$.

**Fig. 5.19.** Light distribution on the experimental table. Solution provided by GA after 500 generations. Three lights sources are positioned in the same growth chamber. $\phi = 13.9 \times 10^{-3}$.

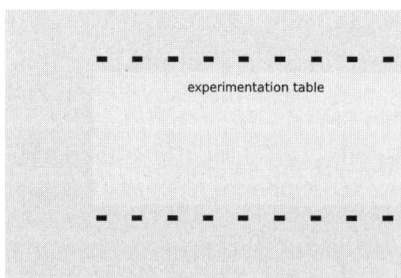

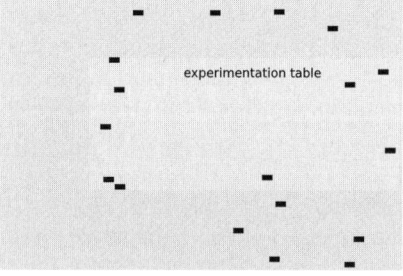

**Fig. 5.20.** The left part scheme provides the real location of the 18 light bulbs in the strader chamber. The right scheme illustrates the solution generated by the GA.

### 5.5.5   Computation Time

The most important part of time consumption is the evaluation of an individual. The other part of the genetic algorithm (cross-over, mutation, ... ) is insignificant. The computation of an individual fitness occupies about 1 second with an Intel T2400, 1.83 GHz Processor. According to our experimental protocol (500 generations of 20 individuals) this requires more than 3 hours of computation time for the search of any solution. This computation time can be increased according to the geometrical complexity of the scene (for example by adding plants on the experimentation table) and/or by scaling the number of photons sent by any light source.

## 5.6  Discussion

We have shown that a GA is able to provide good answers to the positioning problem of light sources in growth chambers. More generally evolutionary methods seem to be an efficient approach for selecting light property in order to solve the inverse problem in lighting [14]. One of the main interest of this approach is the possibility to improve an existent solution. In our first experiments individuals of the initial population are randomly chosen. But it is clearly possible to define the initial locations from existing growth chambers with they own lighting configurations. The GA will then be used for enhancing the light sources positions. Furthermore, by using the elitism property, we can ensure that the new solution will be at the least equivalent to the initial configuration.

The results we outlined are preliminary. They should be completed by comparisons to other methods : expert positioning of lights, classical optimization methods, ant colony optimization [19]. Furthermore this first work made important simplifications both on reflection and emission properties and light geometry and orientation. Extending the approach to more complex (and more realistic) properties is under investigation. Finally the computing time should be reduced in order to provide a useful tool for final users.

## Acknowledgments

Thanks to Quentin Dezetter who developed a part of this project's code during his engineer work placement and to the anonymous reviewers for their helpful comments and remarks.

The software for this work used the GAlib genetic algorithm package, written by Matthew Wall at the Massachusetts Institute of Technology.

## References

1. Boonen, C., Samson, R., Janssens, K., Pien, H., Lemeur, R., Berckmans, D.: Scaling the spatial distribution of photosynthesis from leaf to canopy in a plant growth chamber. Ecological Modelling 156, 201–212 (2002)
2. Cavazzoni, J., Volk, T., Tubiello, F., Monje, O.: Modelling the effect of diffuse light on canopy photosynthesis in controlled environments. In: Acta Horticulturae (2002)

3. Chelle, M., Demirel, M., Renaud, C.: Towards a 3d light model for growth chambers using an experiment-assisted design. In: 4th International Workshop on Functional-Structural Plant Models, Montpellier (2004)
4. Chelle, M., Renaud, C., Delepoulle, S., Combes, D.: Modeling light phylloclimate within growth chambers. In: 5th International Workshop on Functional-Structural Plant Models, Napier, NZ (November 2007)
5. Costa, L., Oliveira, P.: An evolution strategy for multiobjective optimization (2002)
6. De Jong, K.: An analysis of the behaviour of a class of genetic adaptive systems. PhD thesis, University of Michigan (1975)
7. Ferentinos, K.P., Albright, L.D.: Optimal design of plant lighting system by genetic algorithms. Engineering Applications of Artificial Intelligence (2005)
8. Gen, M., Cheng, R.: Genetic Algorithms and Engineering Design. Wiley-Interscience, Chichester (1997)
9. Goldberg, D.E.: Genetic Algorithms in Search, Optimization and Machine Learning. Addison-Wesley Longman Publishing Co., Inc., Boston (1989)
10. Goldberg, D.E.: From genetic and evolutionary optimization to the design of conceptual machines. Evolutionary Optimization 1(1), 1–12 (1999)
11. Holland, J.H.: Report of the systems analysis research group sys. (1975)
12. Jensen, H.W.: Global Illumination Using Photon Maps. In: Proceedings of the Seventh Eurographics Workshop on Rendering Techniques 1996, pp. 21–30. Springer, New York (1996)
13. Jensen, H.W.: Realistic Image Synthesis Using Photon Mapping. A. K. Peters LTD, Natick, Massachussets (2001) ISBN 1-56881-147-0
14. Jolivet, V., Plemenos, D., Poulingeas, P.: Inverse direct lighting with a monte carlo method and declarative modelling. In: ICCS 2002: Proceedings of the International Conference on Computational Science-Part II, London, UK, pp. 3–12. Springer, Heidelberg (2002)
15. James, T.: Kajiya. The rendering equation. In: Computer Graphics (Proceedings of SIGGRAPH 1986), pp. 143–150 (August 1986)
16. Koza, J.R.: Genetic Programming: On the Programming of Computers by Means of Natural Selection. MIT Press, Cambridge (1992)
17. Lafortune, E.P., Willems, Y.D.: Bi-directional Path Tracing. In: Santo, H.P. (ed.) Proceedings of Third International Conference on Computational Graphics and Visualization Techniques (Computergraphics 1993), Alvor, Portugal, pp. 145–153 (1993)
18. Languénou, E., Bouatouch, K., Chelle, M.: Global illumination in presence of participating media with general properties. In: Proceedings of the Fifth Eurographics Workshop on Rendering Techniques 1994, pp. 69–85. Springer, New York (1994)
19. Lee, Z.-J., Su, S.-F., Chuang, C.-C., Liu, K.-H.: Genetic algorithm with ant colony optimization (ga-aco) for multiple sequence alignment. Appl. Soft Comput. 8(1), 55–78 (2008)
20. Measures, M., Weinberger, P., Baer, H.: Variability of plant growth within controlled-environment chambers as related to temperature and light distribution. Canadian Journal of Plant Science 53 (1973)
21. Vafaie, H., Imam, I.: feature selection methods: Genetic algorithms vs greedy-like search. In: Proceedings of the International Conference on Fuzzy and Intelligent Control Systems (1994)
22. Veach, E., Leonidas, J.: Guibas. bidirectionnal estimators for light transport. In: 5th Eurographics Workshop on Rendering, pp. 147–162, juin (1994)
23. Veach, E., Guibas, L.J.: Metropolis light transport. In: 31st Annual Conference Series on Computer Graphics, pp. 65–76 (1997)

# Constructive Path Planning for Natural Phenomena Modeling

Ling Xu and David Mould

Department of Computer Science, University of Saskatchewan, Canada

**Summary.** Path planning is a problem much studied in the context of artificial intelligence, with many applications in robotics, intelligent transport systems, and computer games. In this paper, we introduce the term *constructive path planning* to describe the use of path planning to create geometric models. The basic algorithm involves finding least-cost paths through a randomly weighted regular lattice. The resulting paths have characteristics in common with plants and other natural phenomena; visible structure is imposed on the randomness by the optimization process. This paper explores different arrangements of graph weights and shows the effectiveness of the technique in two detailed examples of procedural models, one for elm trees and one for lightning.

## 6.1 Introduction

The virtual worlds of computer games and computer-animated films are filled with objects both familiar and fantastical. Because of the difficulty of creating sufficiently detailed and numerous models, algorithmic (procedural) techniques have been sought, with the aim of easing the burden on digital artists. In particular, procedural techniques to create models of natural phenomena such as trees, mountains, and clouds have been a subject of long-standing interest by the computer graphics community. Numerous algorithms have been devised for procedurally creating such models [4].

In this paper, we describe an approach to natural phenomena modeling which we term "constructive path planning". The technique involves finding least-cost paths through weighted graphs; we introduced the main idea earlier [16] and here extend and further refine it. The technique involves creating a regular lattice (either a square lattice in 2D or a cubic lattice in 3D) and placing random weights on the edges, then planning least-cost paths through the lattice. By planning paths from a single root node to destination nodes elsewhere in the lattice, a "dendrite" could be created – a sparse, acyclic subset of the original graph. We previously suggested that the algorithm could be used to produce a variety of natural phenomena, and gave examples of coral, lichens, rocks, and lightning.

We make three main contributions in this paper. First, we describe how to make use of the weights in the graph to vary the structure of the output model. Second, we identify the cost value as a useful input to the model and show how cost values can be sensibly used to inform the model appearance. Third, we apply our technique to two specific natural phenomena, elm trees and lightning, and show images and models arising from this application. The lightning we present in this paper is a considerable improvement over previous results.

D. Plemenos, G. Miaoulis (Eds.): Arti. Intel. Techn. for Comp. Graph., SCI 159, pp. 83–102.
springerlink.com ⓒ Springer-Verlag Berlin Heidelberg 2009

The remainder of this paper is organized in four parts. First, we discuss some previous work in natural phenomena modeling. Second, we describe the algorithm of constructive path planning and give details on different edge weight distributions that are useful for modeling purposes. Third, we show results, in the form of images and renderings of our models, and discuss the advantages and disadvantages of constructive path planning. Fourth, we conclude with some recommendations for future work.

## 6.2    Previous Work

Scientific models for natural phenomena abound [1]. In computer graphics, procedural natural phenomena has been studied considerably, with numerous algorithms for procedural terrain, trees, clouds, and other natural objects [4]. The most prevalent method for procedural trees and plants is the replacement grammar L-systems [6]. Diffusion-limited aggregation [15] is a physically motivated model that is also sometimes used. More recently, image-based systems for tree modeling have appeared [9, 14].

L-systems possess two parts: a grammar, with rules describing how tokens are transformed into other tokens or sequences of tokens, and a modeling system, describing how to interpret the tokens as geometric shapes or transforms. Research into L-systems has concentrated on the first part, since rich structures represented by strings of tokens can easily be interpreted into meaningful geometric shapes (perhaps the most common interpretation is the "turtle language", where the tokens are commands directing the turtle to move forward, move backward, or turn in different directions). L-systems, augmented by extensions such as stochastic L-systems, open L-systems, and environmentally-sensitive L-systems, have been quite successful at plant modeling and even modeling of entire plant ecosystems [10, 8, 2]. However, design of replacement rules is challenging.

Unlike L-system focusing on the plant development process, some computer graphics researchers pay more attention to simulating the plant appearance. Image-based methods are used recently to create models for trees [9, 14]. They use photographs taken from the real world to drive the reconstruction of 3D models. The premise of photographs helps to produce very realistic results but also limits the application of modeling specific objects not existing in the real world.

Procedural methods for creating lightning have previously been reported in the literature, notably by Reed and Wyvill [11] and by Kim and Lin [5]. Reed and Wyvill used an ad-hoc particle tracing method to generate a lightning structure, and rendered lightning glow using implicit surfaces. Kim and Lin implemented the dielectric breakdown model for a more physically based approach to lightning; their method produces high quality lightning models and images, but at considerable computational expense.

We previously [16] presented path planning as a modeling technique, and showed how to build a variety of dendritic structures with it, including lightning. We used Dijkstra's algorithm [3] to obtain least cost paths through a random graph, relying on the property that the graphs had positive weights to avoid cycles. Path planning can also be used as a stylization technique for non-photorealistic rendering and modeling, as shown by Long and Mould [7].

While we suggested that the edge weights can be used to control the resulting structure, we did not demonstrate the results of attempting to do so. Also, we omitted to exploit one of the chief pieces of information provided by the algorithm: the cost information at each node. (Cost information was used to determine the path, but ignored once the paths had been found.) We show some examples of controlling the model shape by spatially varying the edge weights, and show how path cost can help to shape our models. Our techniques are employed to create two different types of natural phenomena: elm trees and lightning.

## 6.3  Algorithm

The basic algorithm we employ is that of least-cost paths through a weighted graph [16], where the paths themselves are the modeling primitives. As the extension to the basic algorithm, in this section, we first show how to vary the graph weights spatially to get different effects, and then show how to modify the structures along the path length to produce realistic-looking lightning.

The basic algorithm can be decomposed into the following steps:

1. Create a regular square lattice of nodes: 4-connected in 2D, 6-connected in 3D.
2. Choose edge weights for the graph.
3. Choose a node to be the root of the structure.
4. Apply Dijkstra's algorithm to find path costs to all nodes from the root.
5. Choose path endpoints.
6. Use a greedy algorithm to backtrack from the endpoints to the root, giving the paths.
7. Render the paths.

The collection of the resulting paths form the dendritic structure with the properties of erratic individual paths and bifurcating paths emerging from a common root (as shown in Figure 6.1), where the former property comes from irregular path costs (edge weights) and the latter is achieved by placing proper endpoints and a single root node.

Figure 6.2 shows a few different structures obtained by placing the root at the bottom of the image, scattering endpoints through the graph, and planning paths. The

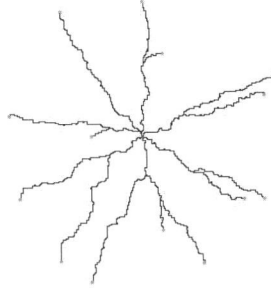

**Fig. 6.1.** A dendritic structure obtained by the basic algorithm of path planning

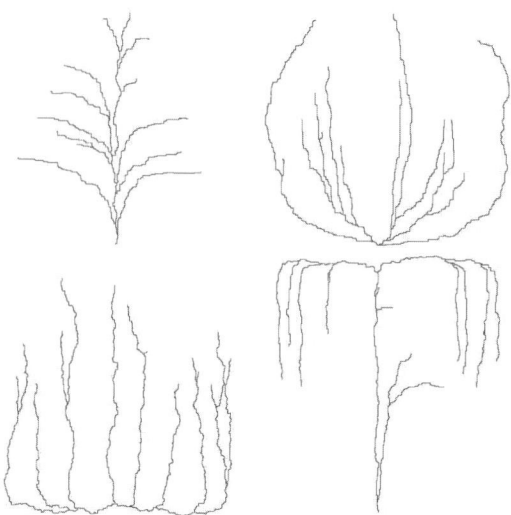

**Fig. 6.2.** Different structures created by varying the edge weights

differences between the structures come from substantial changes made to the distributions of edge weights.

Typically, edge weights were chosen at random. Here, however, we add structured values to the initial random values. Some randomness is needed so that paths do not appear too regular.

In the structure in the upper left of Figure 6.2, edges become more expensive the greater their horizontal distance from the image centre. The result is a structure where the paths hug the central region as much as possible before striking out towards the endpoints. Conversely, in the upper right structure, costs are larger near the image's centre, making it preferable for the paths to bend outward into the cheaper regions before being forced to return for the endpoints. The lower image pair shows a similar contrast: in the lower left, edges are cheaper the closer to the top they are, while in the lower right, edges are cheaper at the bottom. The result is two contrasting structures, the former more tree-like, the latter more like a shrub.

The examples shown involve simple functions of location modifying edge cost, and were done on flat two-dimensional graphs. Such 2D path planning exercises can be completed quickly (in at most a few seconds on a modern desktop computer with modest hardware) and can serve as prototypes for more elaborate models that will be created in 3D. Some 3D models based on these prototypes are shown later in this paper, in Figure 6.7.

The previous examples showed how spatially varying changes to edge weights can affect the structure's overall shape. We now turn to an example of changing the edge weight distribution globally. Previously, we had had edge weights drawn from a uniform distribution (1,max). Here, we suggest that for each edge weight, its value $e$ should be

$$e = R^{\alpha}, \tag{6.1}$$

**Fig. 6.3.** Variations in structure achieved by varying the statistical distribution of edge weights

where $R$ is a random value drawn uniformly from the range (1,max), and $\alpha$ is a parameter controlling the amount of path variation.

The larger the exponent $\alpha$, the greater the disparity between the cheapest and most expensive edges, and therefore, the greater the incentive for the path planner to seek paths consisting of cheap edges. It is no longer profitable to seek short cuts through an expensive edge if many cheaper edges could be used instead. For example, for $\alpha > 1$, when $a + d + c + d = e$, $a^\alpha + b^\alpha + c^\alpha + d^\alpha < e^\alpha$. Thus, increasing $\alpha$ will make it more attractive to take paths with more edges, if those edges are individually cheap. Higher $\alpha$ will result in structures with longer, more roundabout paths through the graph than the structures made with lower $\alpha$. The difference is exemplified by the structures in Figure 6.3, which show results for $\alpha = 0.3$, $\alpha = 1.0$, and $\alpha = 3.0$. The $\alpha = 0.3$ structure has very direct branches, nearly straight lines; the differences between different edge costs have been suppressed. At $\alpha = 1.0$, we have a conventional model, with slight variation in the paths. Finally, with $\alpha = 3.0$, the paths have become more erratic yet, willing to diverge considerably from the Euclidean shortest distance to achieve a better result. Of course, there is no reason that we must stop with $\alpha = 3.0$, and if the desired effect warrants it, still more lengthy and tortuous paths can be obtained with even higher $\alpha$.

In the following subsection, we show how to create plausible-looking lightning. The lightning models use the nonuniform distribution with $\alpha = 3.0$.

### 6.3.1 Lightning

The path planning formulation is already well suited to producing the structure of lightning, simply by placing endpoints and planning paths through the graph. Here, we show how the path costs found by the application of Dijkstra's algorithm can be used to achieve the appearance of lightning as well.

We can produce a visualization of the lightning stroke by assigning a brightness value to each node (pixel) in the graph, computed based on path distance. Let $d$ represent the path cost found by Dijkstra's algorithm. For a pixel at distance $d$, and a thickness factor $T$, we compute a brightness value $V$ as $V = \exp(-(d/T)^2)$. For very small distances, we have $V$ near 1, but the greater the distance, the smaller the $V$. Consequently, pixels (nodes) with greater value of brightness form the wide branch and pixels with small brightness form the slim branch. The pseudocode is shown in Figure 6.5.

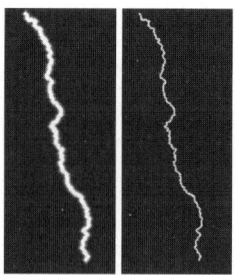

**Fig. 6.4.** Lightning width adjustment. Left: T=5.0; right: T=1.6.

> Input: weighted graph $G$, consisting of nodes $N$ and
> edges $E$; $N_s$, one endpoint of the path; the generator
> $Z$; $d$, the path cost of each node; and $T$, thickness
> factor for the path.
> Output: a brightness value $I(N_i)$ of each node $N_i$.
>
> 1. Apply Dijkstra's algorithm.
> 2. Find the least-cost path $P$ from $N_s$ to $Z$ by greedy
> hill climbing.
> 3. For each node $N_i \in N$:
> if $N_i \in P$, append $N_i$ to $Z$ and set $d$ to 0;
> otherwise, set $d$ to Max.
> 4. Apply Dijkstra's algorithm. Each node $N_i$ has
> an updated cost $d$ and a thickness factor $T$, then
> calculate $V = \exp(-(d/T)^2)$ .

**Fig. 6.5.** Pseudocode for getting a path with a certain width

Two lightning strokes are shown in Figure 6.4, a wider one ($T = 5.0$) and a narrower one ($T = 1.6$). Note that the thicker branch has a weak halo around it, obtained without additional computation. The visible halo around light sources is characteristic of participating media, and lightning typically occurs in stormy conditions where participating media (clouds, rain) are also present.

The previous figure showed how to get lightning strokes of constant width. We can straightforwardly obtain tapering strokes by modulating $T$ along the length of the stroke. We use the original path cost value to do this: we set $T_i$ at a point $i$ along the path to

$$T_i = T_s * (1 - (d_i/d_{max})^\beta) + T_f * (d_i/d_{max})^\beta, \qquad (6.2)$$

for a starting width of $T_s$ and a final width $T_f$ and a path cost value $d_i$ at point $i$. The parameter $\beta$ governs the shape of the tapering; for the lightning, we used $\beta = 1$. Recall that $d_i$ is path cost; in the randomly weighted graphs we used, path cost does not correlate perfectly with edge count, and by using path cost we obtain some additional

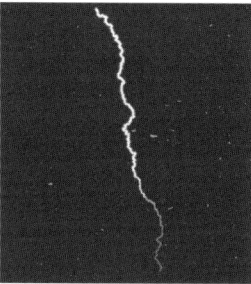

**Fig. 6.6.** Lightning stroke tapering

structure (a nonuniformly but monotonically increasing function) without any further effort. The ability of equation 6.2 to produce tapering is shown in Figure 6.6.

Having completed our description of the algorithms involved in creating our models, we next show some more elaborate results, beginning with 3D structures and elm trees. Our completed lightning images are at the end of the following section.

## 6.4  Results and Discussion

This section contains the more elaborate results of applying constructive path planning to procedural 3D modeling tasks. We first show some somewhat abstract models obtained simply by extrapolating the 2D models seen in section 6.3 to 3D; we then describe how to use our system to generate structures resembling elm trees; and we lastly show how to apply the algorithms of section 6.3 to create realistic-looking lightning.

Figure 6.7 shows different 3D models. These variations were achieved after prototyping the edge weight distribution in 2D, and then performing virtually the same calculation in a 3D space (edge weights that formerly depended on horizontal distance $x$ now depend on horizontal distance $\sqrt{x^2 + y^2}$). These shapes were not intended to resemble any particular type of structure, but they are reminiscent of some kinds of plants, or(in the case of the lowermost example) perhaps not only a tree but also some kind of fantastical, Seussian antler.

We can vary the structure of the objects merely by creating objects with more endpoints. Since the endpoints are randomly placed, and the greedy path traversal is the least expensive part of the algorithm, the more elaborate structures with more endpoints do not actually require more effort to create, either human or computational. Figure 6.8 shows the differences between structures as the number of endpoints varies: the structures on the left have 20 randomly placed endpoints, while those on the left have 36 (the same 20 plus 16 more). Figure 6.9 and Figure 6.10 show 3D models with spatial tropisms simulating some common phenomena. Structures in bottom images in Figure 6.9 simulate trees bending in the wind. They were obtained by making the edge costs higher towards one side of the image, and also distributing the endpoints preferentially to one side of the root node. The structure in the right image in Figure 6.10 simulates the finger cactus with upright paths which were obtained by setting edge costs cheaper towards the bottom side.

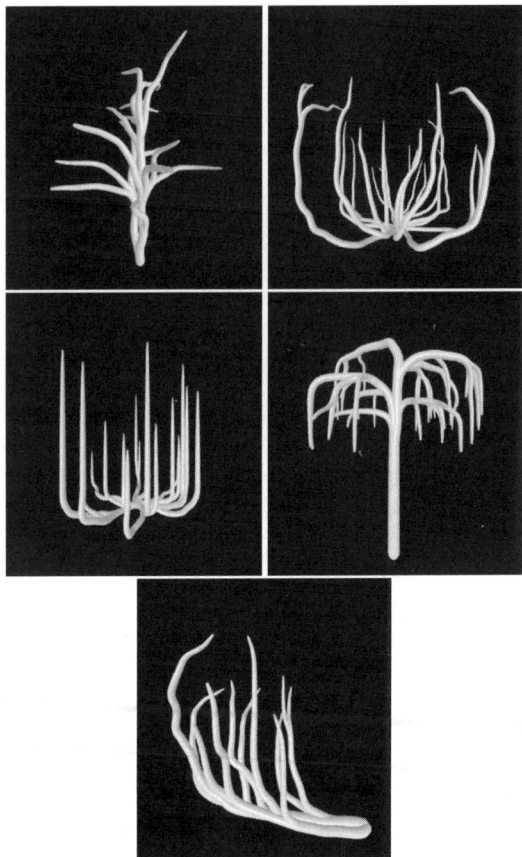

**Fig. 6.7.** Some 3D objects made with different spatial distributions of edge weights

The irregular, winding, forking branches of the structures can be made to resemble elm trees with little difficulty. In Figure 6.11 we show two examples of elm tree models, and in Figure 6.12 we show a third synthetic tree paired with a photograph of a real tree. The two structures have considerable in common, notably the long, curving individual branches, the lack of a single straight trunk, and the spreading of branches away from the base into a wide space above it. Tree models in Figure 6.11 and Figure 6.12 are created within a same framework. We can notice that each of these elm tree models looks different but shares some common characteristics at the same time. The variations come from different edge weights in the 3D lattice and different endpoint positions, while the common places come from the same rules of choosing endpoints and setting path properties (such as tapering length, branch thickness and number of paths). We can also notice that from tree models in Figure 6.13. Figure 6.13 shows the models of trees with a single main trunk. To build this model, we use one path as the main trunk and make slim side branches attach to it.

**Fig. 6.8.** Trees with few endpoints (left) or many endpoints (right)

**Fig. 6.9.** Upper: trees bending in the wind; Lower: our tree models

**Fig. 6.10.** Right: photograph of a finger cactus; left: our 3D model

These trees were created on a relatively low-resolution lattice ($80^3$), making compute times low; computing the structures required only about 4 seconds each. The trees were rendered in Pixie, with tree geometry created by placing a sphere at each path node. Sphere radii were determined using the tapering formulation given in equation 6.2, with $\beta = 0.3$ used for the larger main branches and $\beta = 0.2$ used for the smaller branches, since the tapering of main branches is more obvious than that of side branches.

As an extension of modeling elm trees, we are interested in the behavior of space competition between trees. Competition for space is a common phenomenon that occurs through interactions among immediate neighbors and each individual often suppresses the growth of its neighbors to obtain a disproportionate share of the contested space [13, 12]. Figure 6.14 shows our simulation of space competition between two elm trees. Although the dendrites do not actually communicate with each other during the path planning process, the avoidance comes from the process of setting each node with the least path costs to their nearest generator by Dijkstra's algorithm.

Figure 6.15 shows another example of constructive path planning for modeling natural dendrite. The DLA dendrite (as shown in the left image) is formed by aggregation of particles released far away and experienced a random walk due to Brownian motion. The efficiency is a big problem for DLA. We use the path planning algorithm to generate a dendrite shown in the right image. Besides the similar result, our algorithm shows the dominant advantage over DLA in timing. For creating the shown dendrite, DLA process needs about 8 minutes while our path planning algorithm only needs less than 8 seconds.

Figure 6.16 shows the creation of a lightning structure. This lightning was made in two parts, using two separate 2D graphs with different random edge weights. The first graph contains the main branch and some additional side branches; the second graph contains only side branches. The two structures are composed to form an overall structure, which has richer detail than could be easily produced in a single graph. Also, the paths in the overall structure cross over, giving the impression of a 3D structure

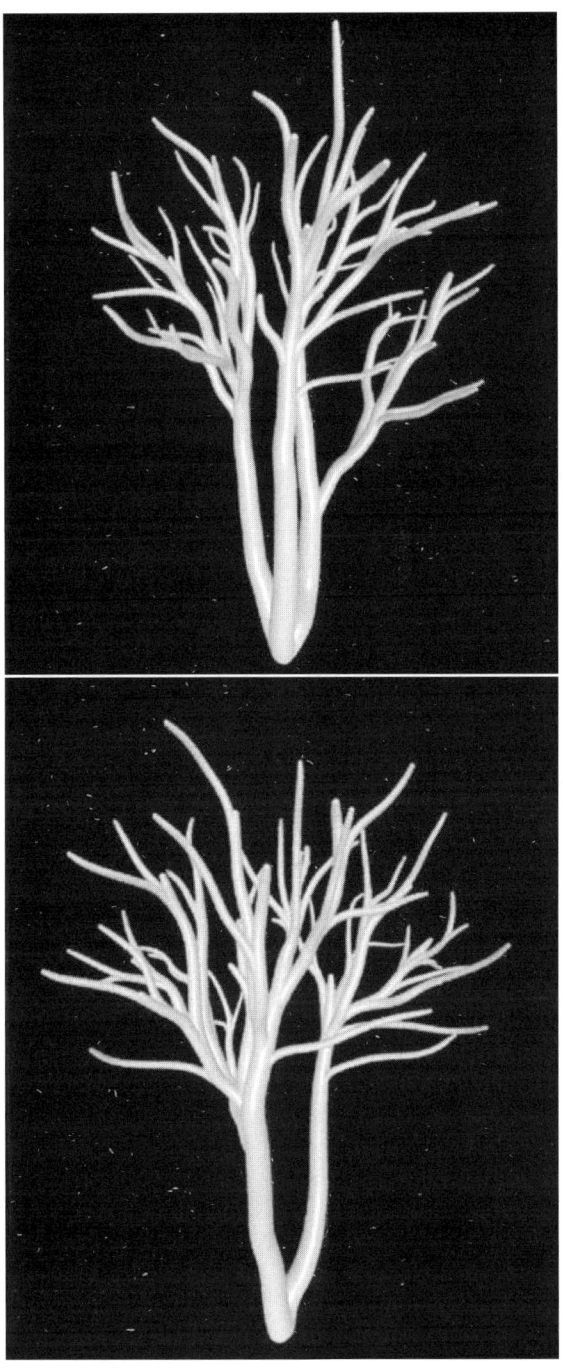

**Fig. 6.11.** Some synthetic elm trees from different random configurations (endpoint placements and edge weights)

**Fig. 6.12.** Comparison between synthetic and real elm tree

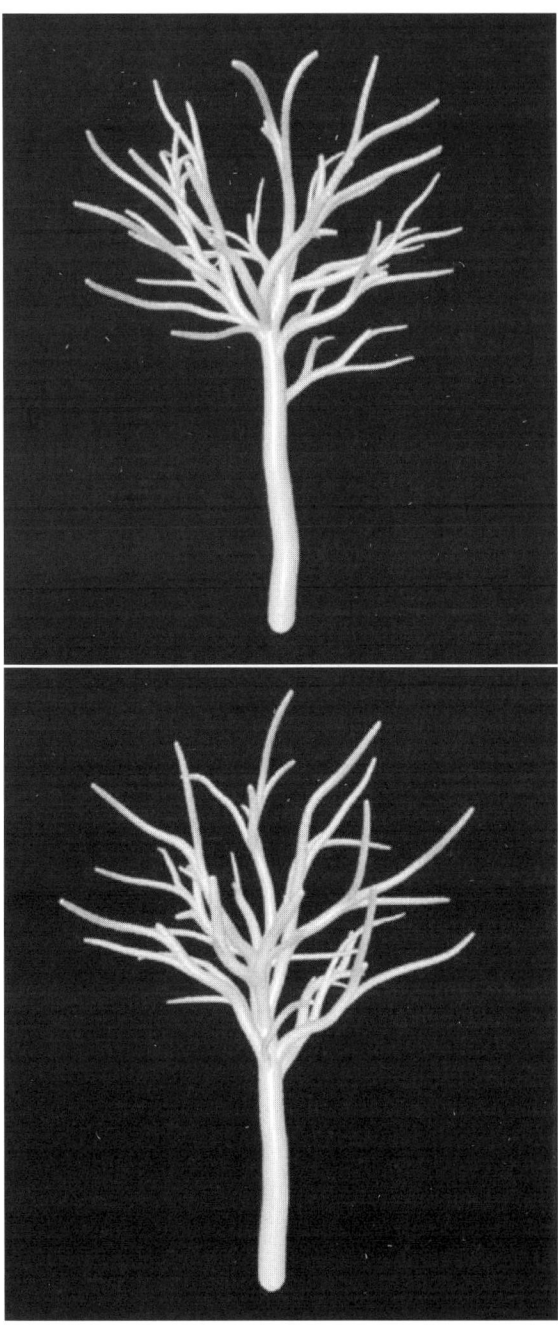

**Fig. 6.13.** Models of trees with a single main trunk

**Fig. 6.14.** Imitated competition behavior between two elm trees

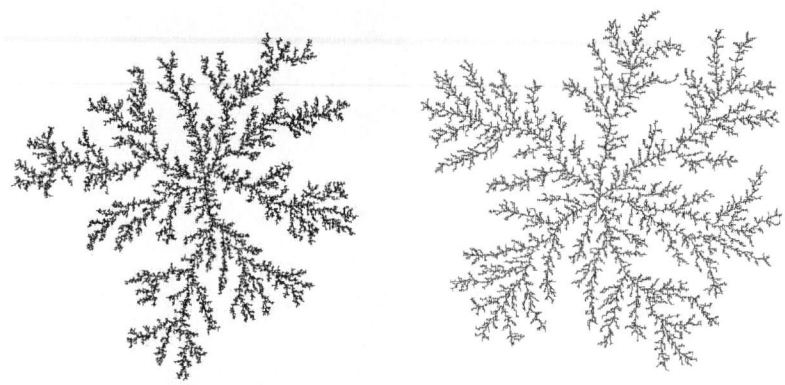

**Fig. 6.15.** Dendrite generated by path planning compared with a DLA dendrite. Left: a DLA dendrite; right: our simulated dendrite.

projected onto a plane. This same approach could be used to produce billboarded trees, if desired.

Figure 6.17 shows a comparison between our path planned lightning and the lightning simulation of Kim and Lin. Although Kim and Lin use a more sophisticated rendering technique to portray their lightning model, we contend that our lightweight system produces equally convincing lightning structure. Note that our system required less than half a minute of computer time to create the model shown, while Kim and Lin report simulation time of several minutes. This means that our system is almost an

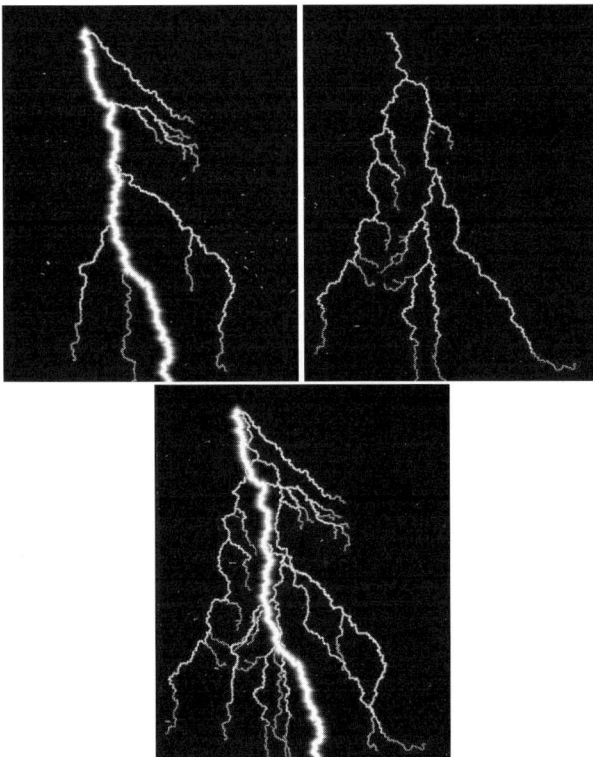

**Fig. 6.16.** Creation of lightning structure. Above left: the main branch and some forks; Above right: additional minor forks. Below: the completed lightning made by composing the two pieces.

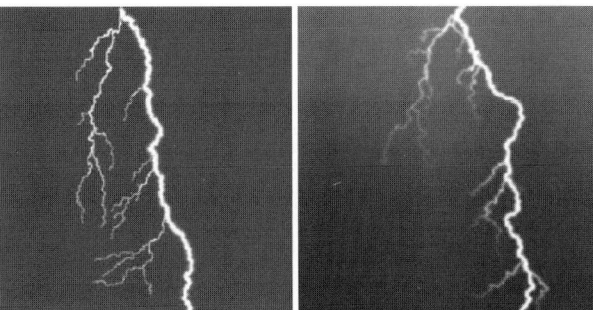

**Fig. 6.17.** Comparison of lightning models. Left: path planned lightning; right: physically based simulation by Kim and Lin.

order of magnitude faster. Compared with our former result, the lightning model from our extended technique looks more realistic (see Figure 6.18). The differences exist in the winding tapering individual branches and different path widths.

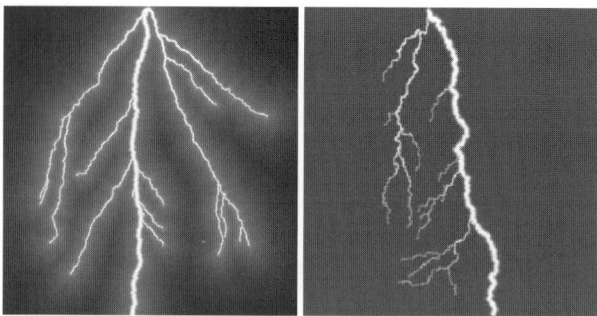

**Fig. 6.18.** Left: lightning produced by our 2007 technique. Right: lightning from our extended technique.

**Fig. 6.19.** Above: Path-planned lightning composited with a photograph of a stormy sky. Below: photograph of lightning.

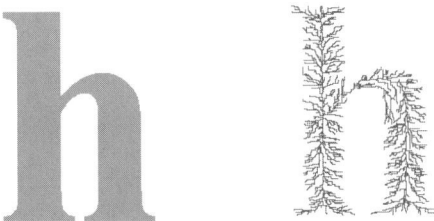

**Fig. 6.20.** Left: a letter shape; right: dendritic letter according to the given shape

**Fig. 6.21.** "hello" written with dendrites

Figure 6.19 shows a comparison with real lightning, where we have taken the lightning structure shown in Figure 6.16 and composed it with a photograph of a stormy sky. The result is convincing; anecdotally, people shown the photographs side by side were only able to identify the synthetic image after studying the two carefully (the main difference is in the halo around the main branch, largely absent in our rendering, and not present in Kim and Lin's either).

However, the main structural characteristics of lightning are represented in our final image. There is a single main lightning stroke, with multiple side branches. The main branch is thicker and has a fairly constant width, while the side branches taper visibly. The lightning is forked, both in having side branches leaving the main branch and in having side branches split from one another. Lastly, owing to our use of multiple (as few as two, in the example shown) parallel 2D graphs stacked together, we can present the strong illusion of a 3D effect.

Figure 6.20 and Figure 6.21 demonstrate the application of our algorithm for creating lichen-writing similar to the result described by Desbenoit et al., reproduced in Figure 6.22. They created the dendritic letters by distributing seeds for DLA in letter shaped areas painted by the designer. Though they did not give the timing figures, we know the open DLA algorithm is very slow. In our method, given a stylized letter shape, we create a graph for the letter and set nodes out of the letter shape illegal for path planning. Then we choose a node as the root and a random set of nodes within the letter-shaped region as endpoints. By finding paths using our path planning algorithm, the resulting paths fill the letter shape in the graph and cause the letter to become visible. Our dendritic letter is comparable to the lichen letter of Desbenoit et al., but observed in details our letter shows more consistence in different parts within the letter

**Fig. 6.22.** Lichen letters created by Desbenoit et al.

**Table 6.1.** Modeling time results

| Model | lattice | endpoints | time |
|---|---|---|---|
| Simple 2D dendrite | $600^2$ | 15 | 0.94s |
| Fractal 2D dendrite | $512^2$ | 8930 | 7.55s |
| Lightning | $600^2$ | 24 | 4.16s |
| 3D trees | $80^3$ | 17 | 3.65s |

shape. It took less than one second to create a dendritic letter composed of 600 paths within a $300 \times 300$ lattice.

Table 6.1 shows modeling time results from our method. The timing results are given for a 1.8GHz P4 with 512 MB RAM. Here the modeling time refers to the time needed to create the lattice and produce the dendritic shape. For 3D tree models shown, besides the modeling time, the average rendering time(the time needed to build spheres for each node in the paths) is less than 4 seconds.

The constructive path planning method has both advantages and disadvantages. The best aspect of the approach is the speed with which coherent structures can be generated, using a minimal (or no) user input. The main disadvantage is the large memory requirement for storing the graph. In this paper, we have shown how spatially organized adjustments to the edge weights can generate various models, and we have shown how to employ constructive path planning to create convincing elm tree models and lightning images. The control afforded by the path planning approach might be useful for some applications; for example, we might want lightning to strike a particular object in the virtual world, and we can place the path planned destination on the desired target.

## 6.5   Conclusion

This paper further explores the capabilities of path planning for model creation. We have shown how constructive path planning can be used to create plausible models of natural objects, specifically lightning and elm trees. The main structural characteristics of these objects can be captured in the path planning framework. Further, we showed how spatially varying the edge weights can produce different organic-looking structures, resembling real-world plant structures or fantastical creatures.

Future work can include applying the method to additional natural phenomena, such as cacti, and increasing the level of user control. One approach would be to permit a user to sketch an object; edge weights could be reduced in the vicinity of the sketch lines, guiding paths towards them.

Since the models we created so far are built in a regular 4-connected in 2D or 6-connected in 3D lattice, as one possible improvement, a non-regular lattice could be used instead of the current regular lattice. By doing so, our algorithm could be applied on the surface of an existing 2D or 3D object and we could generate a smoother dendrite without having to increase the resolution of the graph. To save the cost of building a lattice with high resolution, flexible resolution for different parts of the model could be a promising approach.

Another possibility for future work would be to combine grammars and path planning. Grammars could be used for endpoint distribution, while path planning could still be used to obtain the branches themselves. Users could still adjust the endpoints if desired, but would not be required to, and a suitable grammar might be a good choice for placing endpoints, rather than placing them according to a statistical distribution as we presently do.

## Acknowledgements

Thanks to Jeremy Long and the rest of the IMG group for valuable discussions surrounding path planning for natural phenomena. This work was supported by the University of Saskatchewan, the Canada Foundation for Innovation, and by NSERC RGPIN 299070-04.

## References

1. Ball, P.: The Self-Made Tapestry: Pattern Formation in Nature. Oxford University Press, Oxford (2004)
2. Deussen, O., Hanrahan, P., Lintermann, B., Mech, R., Pharr, M., Prusinkiewicz, P.: Realistic modeling and rendering of plant ecosystems. In: Proceedings of SIGGRAPH 1998, pp. 275–286 (1998)
3. Dijkstra, E.W.: A note on two problems in connexion with graphs. Numerische Mathematic 1(1), 269–271 (1959)
4. Ebert, D.S., Musgrave, F.K., Peachey, D., Perlin, K., Worley, S.: Texturing and Modeling: A Procedural Approach. Morgan Kaufmann Publishers Inc., San Francisco (2003)
5. Kim, T., Lin, M.C.: Physically based animation and rendering of lightning. In: Pacific Conference on Computer Graphics and Applications 2004, pp. 267–275 (2004)
6. Lindenmayer, A., Prusinkiewicz, P.: The Algorithmic Beauty of Plants. Springer, New York (1990)
7. Jeremy Long and David Mould. Dendritic stylization. The Visual Computer, 13 (2008)
8. Mech, R., Prusinkiewicz, P.: Visual models of plants interacting with their environment. In: Proceedings of SIGGRAPH 1996, pp. 397–410. ACM Press, New York (1996)

9. Neubert, B., Franken, T., Deussen, O.: Approximate image-based tree modeling using particle flows. ACM Trans. Graph. 26(3), 88 (2007)
10. Prusinkiewicz, P., James, M., Mech, R.: Synthetic topiary. In: Proceedings of SIGGRAPH 1994, vol. 28, pp. 351–358 (August 1994)
11. Reed, T., Wyvill, B.: Visual simulation of lightning. In: Proceedings of SIGGRAPH 1994, pp. 359–364. ACM Press, New York (1994)
12. Schwinning, S., Weiner, J.: Mechanisms determining the degree of size asymmetry in competition among plants. Oecologia, pp. 447–455 (1998)
13. Slatkin, M., Anderson, D.J.: A Model of Competition for Space. In: Ecology, vol. 65, pp. 1840–1845 (1984)
14. Tan, P., Zeng, G., Wang, J., Kang, S.B., Quan, L.: Image-based tree modeling. ACM Trans. Graph. 26(3), 87 (2007)
15. Witten, T.A., Sander, L.M.: Diffusion-limited aggregation, a kinetic critical phenomenon. Physical Review Letters 47(19), 1400–1403 (1981)
16. Xu, L., Mould, D.: Modeling dendritic shapes - using path planning. In: Braz, J., Vázquez, P.-P., Pereira, J.M. (eds.) GRAPP (GM/R). INSTICC - Institute for Systems and Technologies of Information, Control and Communication, pp. 29–36 (2007)

# 7

## Automatic Generation of Behaviors, Morphologies and Shapes of Virtual Entities

Hervé Luga

VORTEX Research Team, IRIT UMR 5505
Université de Toulouse
UT1, 2 rue du doyen Gabriel Marty 31042 Toulouse, France
herve.luga@univ-tlse1.fr

**Abstract.** This article deals with the use of artificial life paradigms in the field of computer graphics. We show here some clues about the practical use of such techniques that we have developed since 1993 in the VORTEX research team at IRIT. Some samples are given issued from our work dealing with the generation of shapes, behaviors and artificial creatures.

**Keywords:** Virtual reality, Artificial life.

## 7.1 Introduction

Several ways could be followed in the quest to simplify the interactions between users and computers in the field of computer graphics. While some researchers try to increase the power of graphic hardware or to provide automatizations to graphic interfaces, others try to introduce some kind of intelligence inside these virtual environments. This is mainly achieved by the use of paradigms initially developed for artificial intelligence. But if those paradigms have proven their efficiency for solving several problems, they are generally more difficult to use when they have to deal with complex and dynamic environments. At the edge of artificial intelligence one can find some paradigms which took their origins in the observation of how life solves problems. This approach is called "artificial life" and has been introduced mainly by Chris Langton [LANGTON1989].

The vortex team has started in 1993 to use such algorithms to provide the users with more adaptive, responsive and coherent interactions in virtual environments. This chapter shows how apply successfully several artificial life techniques in both the field of modeling and behavioral animation. Its first part deals with some basics about artificial life which focuses on the main paradigms that we use. We then show some applications of this paradigms to shape generation and behavioral simulation. At the end we open new research paths dealing with the automatic generation of complete ecosystems where morphologies and behaviors of artificial creatures emerges directly from their interactions.

## 7.2 Artificial Life

Several definitions of artificial life coexist. Starting from the words of Chris Langton, Artificial Life is *"the study of systems which exhibit some emergent properties"*. This

D. Plemenos, G. Miaoulis (Eds.): Arti. Intel. Techn. for Comp. Graph., SCI 159, pp. 103–121.
springerlink.com

definition means that one can't deduce in a simple way the behavior of an artificial life system by looking at the internal of each of its parts. The key of this definition is the word "simple way" which means that you consider several hierarchical levels in the system's conception. One can know the functioning of each element but, looking at the level where only interactions between these components could be observed, not understanding the global behavior of the whole system.

Artificial life techniques could be classified according to their effects on the underlying system:

- Generators: generators starts usually from a given state and a set of evolutionary rules. The simulation is performed by computing the evolution of the system during time. Generators could also be seen as ontogenetic systems as they act without modification of the system itself. Cellular automata, reaction/diffusion models and L-Systems are some typical examples of generators.
- Evolutionary: evolutionary systems acts on a population of data structures to make it evolves towards an environment pressure. This environment pressure is an abstraction of the problem to solve. Therefore each individual of the population is a "solution trial" which tries to survive by maximizing its utility computed as an adaptation value called fitness. This value is then used to assure the diffusion of its features into a new population. Each loop of this process is usually called "generation". Evolutionary systems acts at a genetic level, modifying the complete structure of the controlled structures. Genetic algorithms [HOLLAND 1975], genetic programming [KOZA 1992], evolutionary strategies are the most known evolutionary algorithms.
- Learning: learning systems modifies its internal structure according to a reward incoming from the environment. Learning systems acts at epigenetic level as they modify an existing system without changing its structure. Classical expert systems, classifier systems [WILSON 1994] and neural networks are learning systems.

In several applications one can mix more than one of these techniques to provide with more complex interactions. For example we use evolutionary algorithms to evolve controllers for cells (which are generators) in an artificial embryogeny application [CUSSAT 2007].

Artificial life techniques have proven their ability to produce systems which exhibit characteristics of living systems. Moreover, they are very efficient in  dynamic environments where they exhibit a lot of abilities as optimization algorithms or to provide with systems able to adapt and generalize knowledge.

## 7.3  Shape Generation

Helping the user in the complex task to design shapes or to manage objects in a 3D scene is a difficult problem. Generally these operations are achieved "by hand" and the knowledge needed to perform those tasks are provided by the designer "know how". According to that, the creation of shapes or the arrangement of 3D scenes is usually a painful and lengthy process.

For helping the designer, one have to provide him with a set of tools able to help that design process. Those tools have to suit the designer  needs while not being to restrictive but to provide a real help. In this part we will show how we apply evolutionary algorithms to solve this problem in two applications: the generation of artistic shapes and as a solver algorithm replacement for declarative modeling.

### 7.3.1  Artistic Generation of Shapes

The aim of this project is to provide artist with a new kind of 3D virtual sculpture tool which enhances creativity [BAILLY SALINS 2007]. Instead of creating virtual clones of real interaction tools and materials such as knives, hammer or rock and clay, we focus on the development of new interaction tools issued from evolutionary algorithms.

Our sculpting tool have to provide the artist with a huge variety of shapes and so don't have to use a too restrictive underlaying definition of those shapes. For that purpose it acts on a  definition of shapes based on implicit surfaces. This definition will allow to describe a wide variety of shapes and has the ability to provide with smooth looking objects.

For proposing  to blend two definition and manipulation levels:

- Mathematical definition of basic shapes: This level provides with a huge flexibility in shape design. These shapes are basics used as bricks which are blended to provide with final shapes.
- CSG-Like manipulation level: This level provides with CSG like tools which could be used to blend shapes issued from the previous level. This provides with a good level of control in shape generation and manipulation.

It is worth to note that each level has already been used in shape optimization applications [JACOB 2001, BEDWELL 1998] with the use of genetic algorithms but have never been blended in the same process for artistic purposes. The links between the two levels are maintained. That means that a CSG shape constructed by blending a shape defined at mathematical level will be modified if the latter is changed.

```
+, -, *, /, mod, round, min, max, abs, expt, log, and,
or, xor, sin, cos, atan, if, dissolve, hsv-to-rgb, vector,
transform-vector, bw-noise, color-noise, warped-bw-noise,
warped-color-noise, blur, band-pass, grad-mag, grad-dir,
bump, ifs, warped-ifs, warp-abs, warp-rel, warp-by-grad.
```

**Fig. 7.1.** Operators used by Karl Sims in its texture generation tool

To help in the design of repetitive or modular structures we propose to uses operators inspired by those proposed by Karl Sims. These operators allows for example to bend a shape or to repeat a structure.

Both levels could use "classic" manipulation algorithms. Those algorithms allow for example union, minus, intersection of shapes. Their utility is to provide with total control over the created objects. To extend these tools we propose to the artist a new kind of tools which are inspired by evolutionary algorithms:

- Crossover: Choosing two shapes, crossover provides with a new set composed by crossing the underlaying tree structure. The user does not have a total control over the arising shapes except the decision to use or not one or more of the provided shapes.

- Mutation: Starting from the definition of one shape provides with a new set created by mutation of the underlaying tree.

This application show us that artificial life paradigm could provide with innovative tools to enhance modeling processes. In opposition the manipulation of such tools is not trivial and only a good experience of the tool make the artist able to efficiently control it's creation.

As an example one can look at shapes from figure 4 which shows a population of seats. This kind of work which in an incursion in the field of evolutionary design needs a lot of expertise and only the artist is able to provide with such knowledge on the system manipulation.

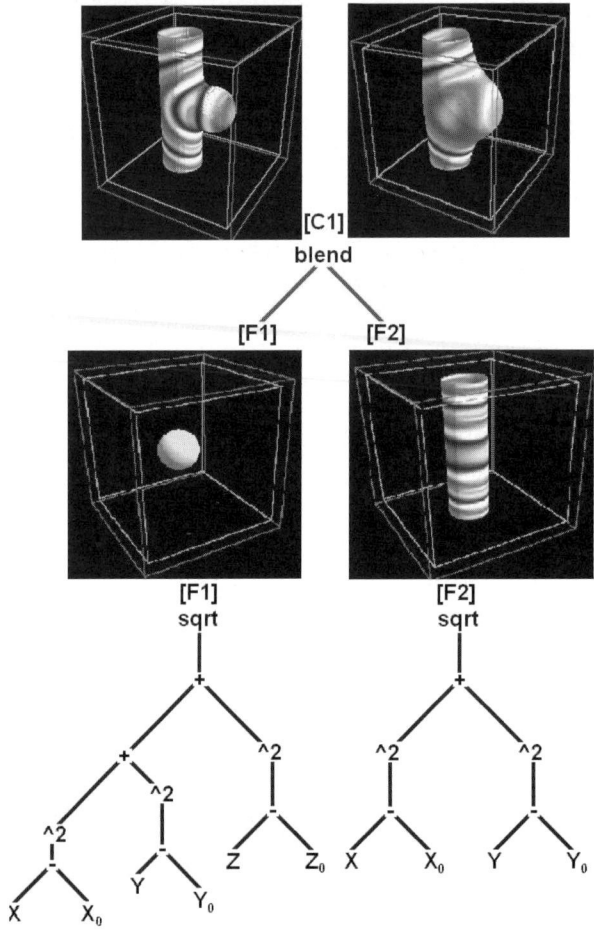

**Fig. 7.2.** Sample definition of a shape

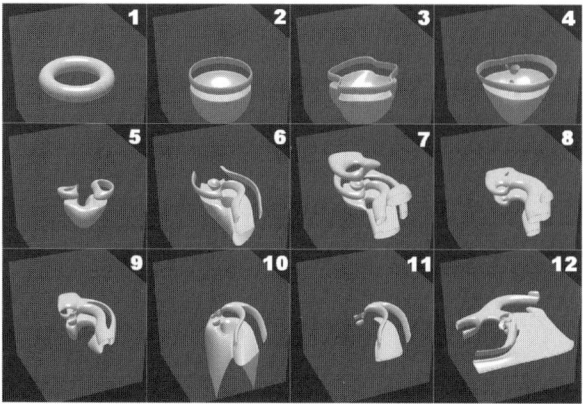

**Fig. 7.3.** A repetitive mutation process starting from one shape

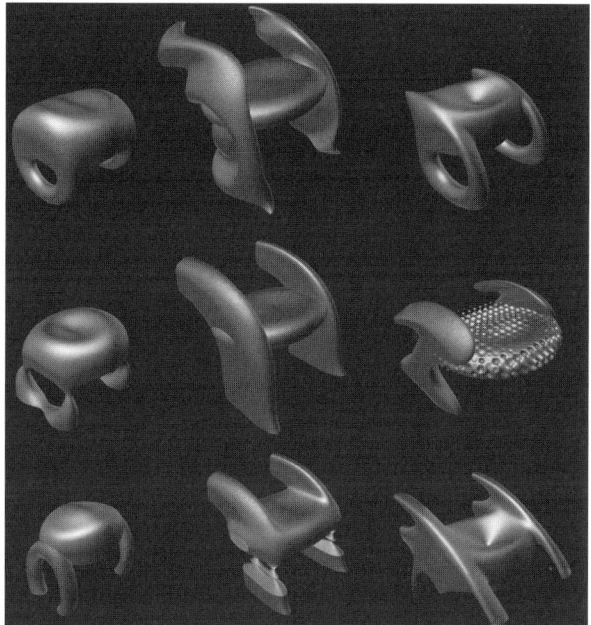

**Fig. 7.4.** A guided use of our tool: generation of a population of seats

### 7.3.2  Constraint Solving in a Declarative Modeling Application

This works [SANCHEZ 2000] shows how one can use artificial evolution as a constraint solver instead of classical CSP trees. The problem consists in the placing of objects in a 3D scene. The objects have to settle according to a set of constraints which could be either external to the objects or relative between several objects.

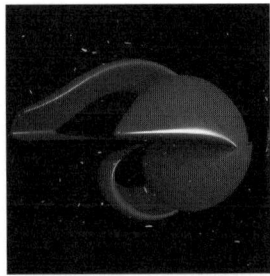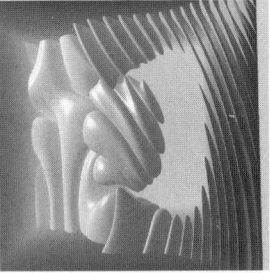

**Fig. 7.5.** Some shapes issued from our modeling tool

The aim of this application is essentially to prove the ability of such evolutionary algorithms to be efficient in that kind of NP problem. In that manner it could be directly linked with our previous works in the constrained object generation using a genetic algorithm. Furthermore we show that it provides with an elegant solution to solve that kind of problems in cases where the system is under or over constrained.

This work is linked with the works about declarative modeling which are developed in our research team. The set of constraints the objects have to satisfy could be explained in a natural way. These constraints are then arranged into a tree which is normally solved using CSP algorithms.

For our application the position and the orientation of each object is defined by a set of real numbers which are linked to the degrees of freedom the object has. Each value is discretized to reduce the state space and ensure the efficiency of our solving

---

**- Chairs 1 and 2:**
Zone INFRONTOF table OR Zone UNDER table
AND Orientation table BETWEEN $3\pi/4$, $5\pi/4$
AND Distance table BETWEEN 90, 120
**- Chairs 3 and 4 :**
Zone RIGHTOF table OR Zone UNDER table
AND Orientation table BETWEEN $\pi/4$, $3\pi/4$
AND Distance table BETWEEN 90, 120
**- Chairs 5 and 6 :**
Zone BEHINDOF table OR Zone UNDER table
AND Orientation table BETWEEN $7\pi/4$, $\pi/4$
AND Distance table BETWEEN 90, 120
**- Chairs 7 and 8 :**
Zone LEFTOF table OR Zone UNDER table
AND Orientation table BETWEEN $5\pi/4$, $7\pi/4$
AND Distance table BETWEEN 90, 120

---

**Fig. 7.6.** A set of constraints to put 8 chairs around a table

algorithm while maintaining a sufficient diversity. Then, each individual of the population is a layout trial which have to be evaluated according to three criteria:

- The collision with static objects: Objects cannot intersect with static objects in the scene.
- The gravity constraint: Objects has to be in equilibrium according to the gravity force.
- The given constraints: Objects have to respect the set of given constraints or to maximize the number of satisfied constraints.

For each object in the solution trial we compute a fitness for each constraints it has to solve. These evaluations are then blended to provide with the global note linked with one of the of the previously cited criterias.

The result of each evaluation is translated into a real number in the [0,1] interval representing the trial performance. The three numbers are then blended to provide with a single evaluation function.

As a result we find that this kind of algorithm is able to work efficiently as a replacement to classical constraint solver algorithm. Using nowadays classical computer

$$\text{If } orientation_{trial} \leq orientation_{requested}$$
$$\text{Then } Fitness_{trial} = \frac{orientation_{g\grave{e}ne}}{orientation_{requested}} \times 100$$
$$\text{Else } Fitness_{trial} = \frac{orientation_{requested}}{orientation_{trial}} \times 100$$

**Fig. 7.7.** Example of fitness linked with the "orientation" constraint

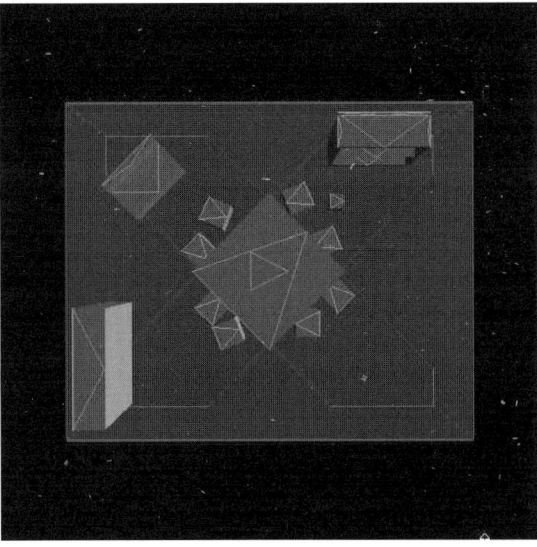

**Fig. 7.8.** One sample solution from the previously defined set of constraints

Generations :248

Computing time/gen :
814 ms

Solver global response
time : 201,872 s
(≈ 3 min 22 s)

**Fig. 7.9.** "Placing 20 books on a shelf" and related computing time

one can find that computing times are generally suited for an interactive use of the solver, and can be compared with those provided using CSP algorithms. Moreover the ability of the algorithm to provide with a huge variety in the proposed solutions is very interesting when dealing with multiple occurrences of the same set of constraints.

One main drawback of this approach is that it is difficult to evaluate when the evolution has to be stopped. For more efficiency we propose to use an interactive interface which shows the evolution at work to the user allowing it to stop or to continue the current evolution process.

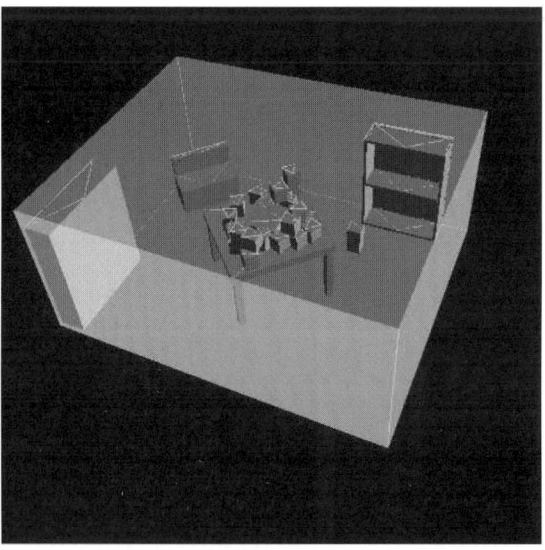

**Fig. 7.10.** Automatic placing of 20 objects on a table but on the box

As opposite one has to notice that knowledge which is used to design both the genotypes and the fitness functions is not trivial. Complexity has moved from the algorithm to the tuning of the solver. We have for example designed several particular genetic operators which took into account the specifics of our application.

## 7.4  Generation of Behaviors

As with modeling, one can think about the application of artificial life algorithms to behavior generation. We develop several studies in that field which covers a wide range of applications starting from the automatic generation of simple moves to the complete joint generation of morphologies and behaviors of virtual entities.

In that section we will show three application samples in that field with the generation of behaviors: The generation of articulated objects positions, the generation of behavioral control modules and finally the generation of artificial creatures.

### 7.4.1  Articulated Objects Animation

This work deals with the generation of the moves of an articulated object which has to follow a target and to take into account a set of environmental constraints [BALET 1997]. The main difference with the work we described in the previous paragraph is that the moves have to be computed in real time. One can describe articulated objects as a set of angles linked with the degrees of freedom of the object. To provide with real time interaction we use a genetic algorithm instead of inverse kinematic which is the more commonly used algorithm in that kind of problems [KWAITER 1998]. Each individual of the population is a vector linked with the changes one has to apply to each angle to compute the next object position.

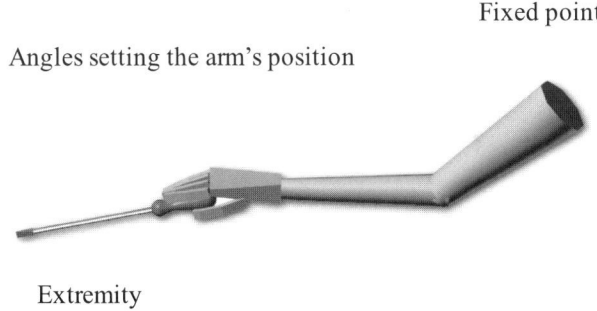

**Fig. 7.11.** An articulated object representing a virtual arm

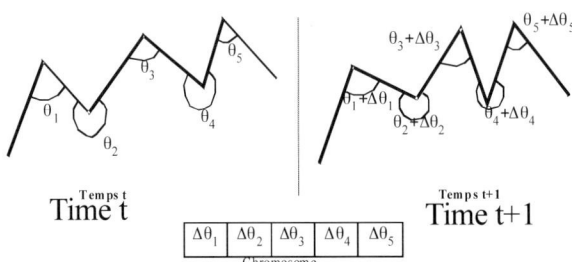

**Fig. 7.12.** Evolution of the angles controlling a wire between two time steps

The fitness function has only to compute the final distance between the extremity of the object and the target point. As opposite to inverse kinematics our algorithm can even provide with solutions when the target point is out of the object's range. In that case it provides with the position which is the closest of the target.

Our experimentations proves the ability of our system to act in real time with 20 degrees of freedom and to provide with efficient solutions. Moreover, adding constraints to the moves of the articulated object like exclusion/preferred zones could be achieved by modifying the fitness function according to the crossing between those zones and the articulated object. It is worth noting that we have developed another version which directly computes the angles to provide with static placement of wires and which is able to deal with more complex constraints such as heat/electromagnetic avoidance.

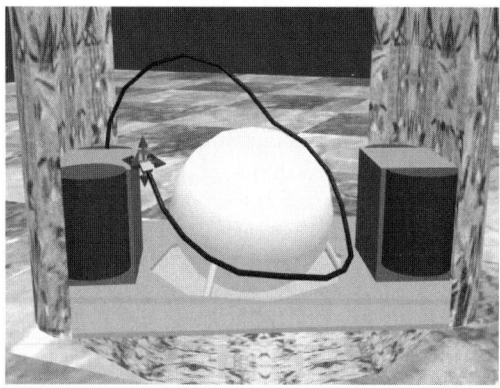

**Fig. 7.13.** A simulation showing a wire avoiding a white tank

This realtime application of genetic algorithms to generate motions has been applied in several simulations as basic motion controller for worm-like entities [PANATIER 2000].

### 7.4.2  Automatic Generation of Behavioral Modules

In the quest to provide virtual human with coherent and effective behaviors we have proposed a behavioral system called VIBES. That system is based on classical behavioral animation paradigms [TERZOPOULOS 1989]:

- Each entity has its own sensors, actuators and behavior.
- An entity can only sense and perform in the environment according to the last cited.
- An entity cannot sense the difference between  the ones controlled by computer and by humans.

The action selection is achieved by a hierarchical behavioral system based on behavioral modules. These behavioral modules are a kind of automatons which are able to compute at every time step the actions to put in an action list that controls the

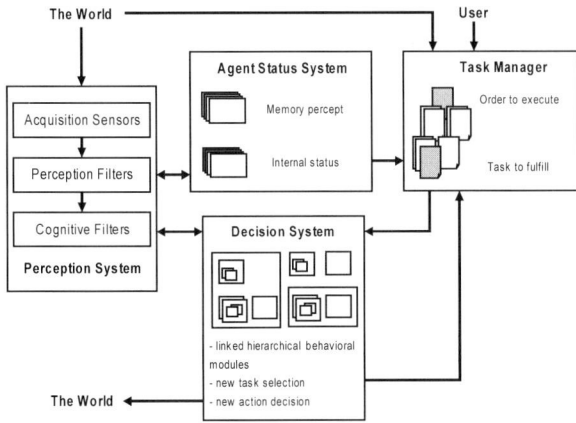

**Fig. 7.14.** An overview of the VIBES behavioral simulation system

actuators of the entity. The entity performs in sequence the actions described in its action list but can rearrange them according to the priority of the entity's goals or to new situations sensed from the environment.

Our aim is to provide the designer with a system able to create automatically such behavioral modules. For that purpose we decide to use classifier systems which is a rule based system where the rules could be discovered and reinforced in the environment.

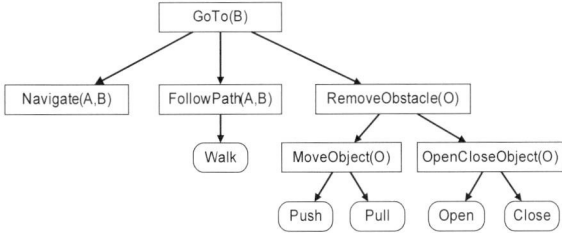

**Fig. 7.15.** A sample behavioral module and its subcomponents : Goto

In a bottom up approach we decide to use this learning system to create the links between several existing behavioral modules to provide with a more complex one. The rules then describe the transitions between the modules according to the state of the entity and the values sensed from the environment.

We choose to apply our proposal to the offline generation of new behavioral modules. Learning is applied by reinforcement. That means that we put the actor which integrates the module to optimize in a simulated environment and evaluate the result of each of its actions to provide a reward to the classifier system. After some simulation steps the rule system converges to a set of condition/action pairs which could be translated as an automaton. This automaton then describes a new basic behavior and could be in turns used to produce more complex ones.

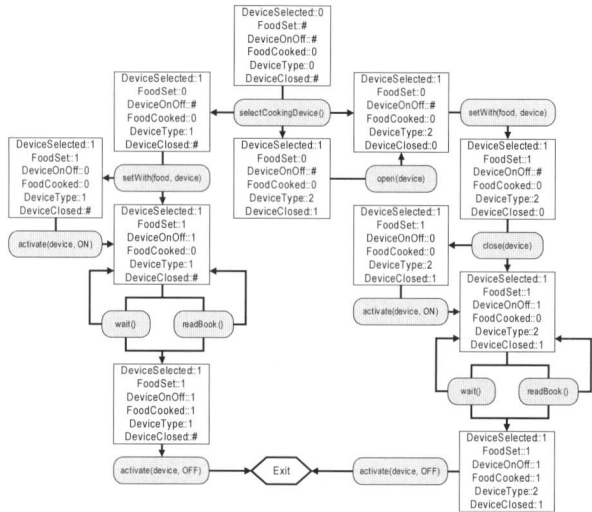

**Fig. 7.16.** An example of generated automaton: cookfood. Behavioral modules are greyed

We successfully apply this approach on the generation of several modules and show the ability of classifier systems to provide with a complex sequence of actions. Moreover we show that our system could be used to integrate new actions into an existing module. An example of such integration could be seen in the previous figure where the actions "wait" and "readbook" appear "in parallel". We also provide some clues to make our system able to learn online by the use of a behavioral consistency control system which can stop an inconsistent action prior to its execution.

### 7.4.3  Conjoined Generation of Morphologies and Behaviors

In the quest to provide with completely evolved virtual actors we have investigated the field of conjoined generation of morphologies and behaviors toward an environment pressure [LASSABE 2007]. This works follows the one of Karl Sims [SIMS 1994] which first exhibit the ability of artificial evolution to generate complete complex creatures.

Virtual creatures are immersed in a 3D environment where physical laws are simulated. Therefore they can interact with objects (and other creatures) in a realistic way. Creature behavior is controlled by setting up links between the sensors of the entity and joints between the parts of its morphology. Joint control is generally realized by the use of functions controlling either the angle, the speed or the torque applied to the joint.

A genetic algorithm is then used to provide with both the morphology and the controllers embedded inside each part of the creature. Authors generally uses some kind of function composition networks for the latter. They also sometimes include some sensors like gyroscopic or contact sensors into these networks [KOMOSINSKI 2000].

Our work aims at extending this work by confronting the creatures to a more complex environment. For that purpose we settle several experimentations where virtual

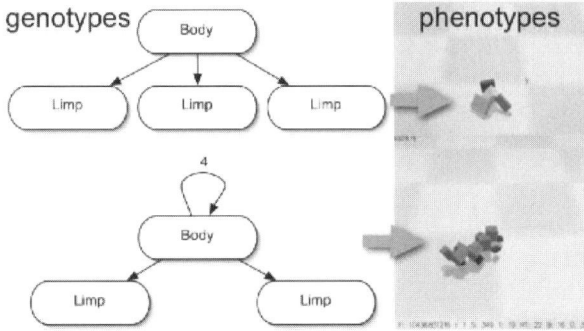

**Fig. 7.17.** Some graftal and the corresponding morphologies

environments integrates several new objects like stairs, boxes, skateboard,... and let the evolution provide creatures which are able to interact with those objects.

The morphology of our creatures uses classical approach proposed by Karl Sims which uses graftals to provide creatures with a compact representation which allows to create a wide variety of shapes. Moreover graftals allows to deal with modularity by the use of recursive rules. On the contrary we propose a new behavioral controller which is based on a kind of classifier system able to provide with complex patterns to control the virtual mussels of the creatures.

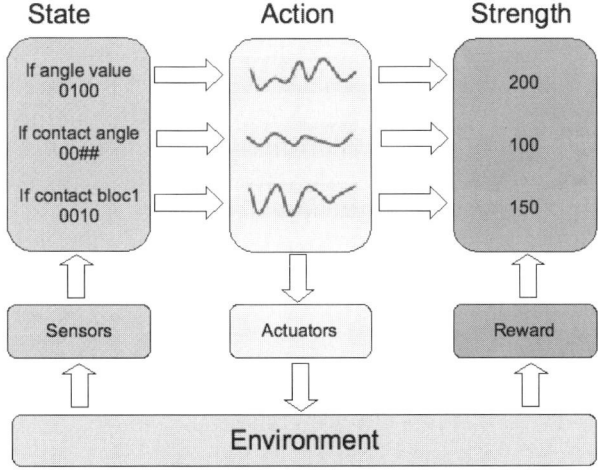

**Fig. 7.18.** Links between the rules, the actions and the environment

At first we start by validating our system with a classical problem: the evolution of creatures which are able to walk. The fitness function is the mean speed of the creature during the 40 seconds of each run. So our evolution process consists in the simulation of each individual in the population which provides with a note reflecting its performance.

**Fig. 7.19.** Sample creatures generated during the validation of our system

We find interesting results which shows the ability of our behavioral system to provide with good control sequences. We also exhibit that each run provides with a set of satisfying results which represents several variations of the same basic behavior and morphology. Providing with diversity is then achieved by making a lot of runs.

Moreover we study the effects of allowing or not the creatures to be explicitly symmetric by changing the fitness function definition. We start by considering only the moves of the creatures without taking into account its direction. We then changed the fitness function to compute the running distance along the "x axis". We find that the latter definition provides with more symmetric creatures.

After doing those experiments to show the validness of our generation system, we then settle more complex environments where creatures has to interact with static objects.

The first benchmark we define consists in adding stairs to the environment. The fitness function is the distance on the stairs the creature has performed during simulation time.

We observe that several evolution strategies emerges to solve this problem. Some creatures aims at develop a huge body which allows to climb easily. Some others uses modularity to create long worm like entities which crawl up the stairs. One can remark that some creatures uses a side effect of our physical simulation by the way of

**Fig. 7.20.** Sample creatures climbing stairs

**Fig. 7.21.** Several creatures moving across boxes

being thrown away to the top of the stairs. This exhibit another useful application of such evolutionary algorithms which is to find some traps into simulation systems.

In another simulation the creatures has to walk thru a set of boxes without falling. One can identify two basic behaviors: Developing morphologies which are larger than the blocs or learning to swarm from one block to the other without falling in the trenches. To exhibit the ability of the creatures to adapt to changes in the environment we modify the size of the blocks and the length of the trenches. This provides with larger creatures or with creatures which are able to lean on a block then to move to the next one.

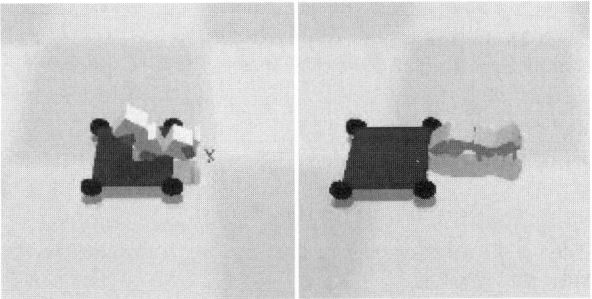

**Fig. 7.22.** Creatures using a skateboard

We then develop another set of simulations where the virtual entities are confronted with mobile objects.

The first object is a skateboard the entity has to move with in the environment. This skateboard is composed of a plank and of four fixed wheels. As a consequence it is not able to turn without being lifted. We define as fitness the distance that the creature rides the skateboard during a run. At the beginning of the simulation the evaluated creature is placed on the board which is still. We find that several strategies emerges. Some creatures tends to push the board then to run towards it. Some others stays on the board and push it with a part of their body acting as a pole.

We show that evolution is able to provide with complete complex creatures evolving in a wide variety of environments. One important point is the predominance of worm like morphologies which are certainly a side effect of our genetic operators and

**Fig. 7.23.** Creatures pushing a block

of the structure of our controller. However the latter shows its ability to be a candidate replacement to the function composition networks which are generally used in that kind of problems. We work now on some applications of this system to the co evolution of creatures. Our final aim is to settle a complete ecosystem where the fitness function is not exogenous but directly linked with the survival of the creature in this environment.

## 7.5   Conclusion

In that chapter we describe several applications of the artificial life paradigms in the field of computer graphics. The variety of the provided examples are a subset of the whole applications we have developed at IRIT since 1993. They show that those techniques are well suited to a wide variety of problems. It is worth noting that the development of such algorithms follows the increase of computing power provided by workstations.

On the outline we can see that nowadays artificial life systems, using mimic of natural processes as basic, are now able to generate artificial creatures which exhibit properties of living organisms. One most impressive example are the autonomous robots of Lipson [LIPSON 2007] which are able to self reproduce, self repair, auto organize. They prove the ability of artificial life algorithms to provide with lifelike automatically generated entities.

We now work on several projects which aims at extending the field of appliance of those algorithms to the complete generation of creatures in ecosystems starting from a single generalized cell. Moreover we continue working on classical behavioral animation by proposing new clues to provide artificial agents with cognitive abilities without having to explicitly represent high level information.

## References

Alexe, A., Barthe, L., Gaildrat, V., Cani, M.P.: Shape Modeling by Sketching using Convolution Surfaces. In: Pacific Graphics, Macau (China) 10/10/05-14/10/05 ACMSIGGRAPH / IEEE (October 2005)

Badler, N., Allbeck, J., Zhao, L., Byun, M.: Representing and Parameterizing Agent Behaviors. In: Proc. Computer Animation, Geneva, Switzerland. IEEE Computer Society, Los Alamitos (2002)

Salins, I.B., Luga, H.: Artistic 3D Object Creation Using Artificial Life Paradigms. In: Butz, A., Fisher, B., Krüger, A., Olivier, P., Owada, S. (eds.) SG 2007. LNCS, vol. 4569, pp. 135–145. Springer, Heidelberg (2007)

Baja, J., Blinn, J., Bloomenthal, J., Cani, M.P., Rockwood, A., Wyvill, A.B., Wyvill, B.G.: Introduction to Implicit Surfaces. Morgan Kaufmann Publishers Inc., San Francisco, CA, USA (1997)

Balet, O., Luga, H., Duthen, Y., Caubet, R., Provis,: A platform for virtual prototyping and maintenance tests. In: Proceedings of IEEE Computer Animation 1997, Geneva, Switzerland, pp. 39–47. IEEE Computer, Los Alamitos (1997)

Bedwell, E.J., Ebert, D.S.: Artificial evolution of implicit surfaces. In: ACM SIGGRAPH 1998 Conference abstracts and applications, Orlando, Florida, United States (1998)

Bentley, P.J.: Evolutionary Design by Computers. Morgan Kaufmann Publishers Inc., San Francisco (1999)

Blumberg, B., Downie, M., Ivanov, Y., Berlin, M., Jonhnson, P., Tomlinson, B.: Integrated Learning for Interactive Synthetic Characters. In: SIGGRAPH 2002, San Antonio, TX, July 21-26 (2002)

Bongard, J., Paul, C.: Investigating morphological symmetry and locomotive efficiency using virtual embodied evolution. In: Mayer, J.-A., et al. (eds.) From Animals to Animats: The Sixth International Conference on the Simulation of Adaptive Behaviourp, 16, 93 (2000)

Bongard, J., Pfeifer, R.: Evolving complete agents using artificial ontogeny. In: Morphofunctional Machines: The New Species, p. 50, 52, 60, 87, pp. 237–258 (2003)

Bongard, J., Pfeifer, R.: How the Body Shapes the Way We Think: A New View of Intelligence. Bradford Books (2007) ISBN-10:0-262-16239-3

Bongard, J., Lipson, H.: Integrated design, deployment and inference for robot ecologies. In: Proceedings of Robosphere, p. 19, 20 (2004)

Bongard, J., Lipson, H.: Once more unto the breach, automated tuning of robot simulation using an inverse evolutionary algorithm. In: Proceedings of the Ninth Int. Conference on Artificial Life (ALIFE IX), p. 19, 110 (2004)

Bret, M.: Virtual living beings. In: Heudin, J.-C. (ed.) VW 2000. LNCS (LNAI), vol. 1834. Springer, Heidelberg (2000)

Brooks, R.: Artificial life and real robots. In: European Conference on Artificial Life, p. 15, 18, pp. 3–10 (1992)

Brooks, R.: How to build complete creatures rather than isolated cognitive simulators. Architectures for Intelligence, p. 49, pp. 225–240 (1991)

Brooks, R.: Intelligence without representation, Artificial Intelligence 47(1-3), p. 4, 139–159 (1991)

Brooks, R., Selman, B., Dean, T., Horvitz, E., Mitchell, T.M., Nilsson, N.J.: Challenge problems for artificial intelligence. In: Thirteenth National Conference on Artificial Intelligence - AAAI, p. 4, pp. 1340–1345 (1996)

Cussat-Blanc, S., Luga, H., Duthen, Y.: Using a single cell to create an entire organ. In: Brooks, T., Ikei, Y., Peterson, E., Haller, M. (eds.) International Conference on Artificial Reality and Telexistence (ICAT 2007), Esbjerg, pp. 300–301. IEEE Computer Society, Los Alamitos (2007)

Darwin, C.: On the origins of species by means of natural selection (1859)

Funge, J., Tu, X., Terzopoulos, D.: Cognitive Modeling: Knowledge, Reasoning and Planning for Intelligent Characters. In: SIGGRAPH 1999, Los Angeles, CA, August 11-13 (1999)

Goldberg, D.E.: Genetic Algorithms in Search, Optimization, and Machine Learning. Addison-Wesley, Reading, p. 11, 14, 27 (1989)

Gritz, L., Hahn, J.K.: Genetic programming evolution of controllers for 3-D character animation. In: Koza, J.R., Deb, K., Dorigo, M., Fogel, D.B., Garzon, M., Iba, H., Riolo, R.L. (eds.) Genetic Programming 1997: Proceedings of the Second Annual Conference, p. 19, pp. 139–146. Morgan Kaufman, San Francisco (1997)

Holland, J.H.: Adaptation in Natural and Artificial Systems. University of Michigan Press, Ann Arbor (1975) (Republished by the MIT Press 1992)

Jacob, C., Kwong, H., Wyvill, B.: Toward the creation of an evolutionary design system for implicit surfaces. In: Western Computer Graphics Symposium, Skigraph 2001, Sun Peaks Resort, British Columbia (2001)

Jones, M.W.: Direct surface rendering of general and genetically bred implicit surfaces. In: 17th ann. conf. of eurographics (uk chapter) cambridge, pp. 37–46 (1999)

Kallmann, M., Monzani, J., Caicedo, A., Thalmann, D.: ACE: A Platform for the Real Time Simulation of Virtual Human Agents. In: EGCAS 2000 - 11th Eurographics Workshop on Animation and Simulation, Interlaken, Switzerland (2000)

Kawaguchi, Y.: Electronic Art and Animation Catalog, pp. 90–91. ACM Press, New York, NY, USA SIGGRAPH 2005: ACM SIGGRAPH (2005)

Komosinski, M.: Framsticks: a platform for modeling, simulating and evolving 3D creatures. In: Artificial Life Models in Software, ch. 2, p. 45, pp. 37–66. Springer, Heidelberg (2005)

Komosinski, M.: The world of framsticks: Simulation, evolution, interaction. In: VW 2000: Proceedings of the Second International Conference on Virtual Worlds, London, UK, p. 15, 16, 17, 42, 45, 64, 87, pp. 214–224. Springer, Heidelberg (2000)

Koza, J.R.: Genetic programming: on the programming of computers by means of natural selection. MIT Press, Cambridge (1992)

Koza, J.R.: The genetic programming paradigm: Genetically breeding populations of computer programs to solve problems. In: Soucek, B., Group, I.R.I.S. (eds.) Dynamic, Genetic, and Chaotic Programming, pp. 203–321. John Wiley, New York (1992)

Kwaiter, G., Gaildrat, V., Caubet, R.: Controlling Objects Natural Behaviors with 3D Declarative Modeler. In: Proceeding of Computer Graphics International, CGI 1998, Hanover, Germany, pp. 248–253, June 24-26 (1998)

Kwaiter, G.: Modélisation déclarative de scènes : étude et réalisation de solveurs de contraintes, Thèse de Doctorat de l'Université Paul Sabatier, Toulouse (December 1998)

Langton, C.G.: Artificial Life: Proceedings of an Interdisciplinary Workshop on the Synthesis and Simulation of Living Systems. Addison-Wesley Longman Publishing Co., Inc., Boston, MA, USA (1989)

Langton, C.G.: Proceedings of the Interdisciplinary Workshop on the Synthesis and Simulation of Living Systems (ALIFE 1987). Santa Fe Institute Studies in the Sciences of Complexity, vol. 6. Addison- Wesley, Los Alamos, p. 15, 18 (1987)

Lamarche, F., Donikian, S.: Automatic Orchestration of Behaviours through the management of Resources and Priority Levels. In: Autonomous Agents and Multi Agent Systems, Bologna, Italy (July 2002)

Larive, M., Gaildrat, V.: Wall Grammar for Building Generation. In: Spencer, S.N. (ed.) Graphite, 4th International Conference on Computer Graphics and Interactive Techniques, South-East Asia, Kuala Lumpur, Malaysia, 29/11/06-02/12/06, pp. 429–438. ACM Press, New York (2006)

Lassabe, N., Luga, H., Duthen, Y.: Evolving creatures in virtual ecosystems. In: Pan, Z., Cheok, A.D., Haller, M., Lau, R.W.H., Saito, H., Liang, R. (eds.) ICAT 2006. LNCS, vol. 4282, pp. 11–20. Springer, Heidelberg (2006)

Lassabe, N., Luga, H., Duthen, Y.: A New Step for Evolving Creatures. In: IEEE Symposium on Artificial Life (IEEE-ALife 2007), Honolulu, Hawaii, pp. 243–251. IEEE, Los Alamitos (2007)

Le Roux, O.: Modélisation déclarative d'environnements virtuels: contribution à l'étude des techniques de génération par contraintes. PhD thesis, Université Paul Sabatier (2003)

Lipson, H.: Curious and creative machines. 50 Years of AI, p. 2, pp. 316– 320 (2007)

Luga, H., Duthen, Y., Pelle, R., Berro, A.: Extended algebraic surfaces generation for volume modeling: an approach through genetic algorithms. In: Proceedings of Visualization and Modeling, Leeds, British Computer Society (1995)

Ngo, J.T., Marks, J.: Spacetime constraints revisited. Computer Graphics 27, p. 28, 343–350 (1993)

Panatier, C., Sanza, C., Duthen, Y.: Adaptive Entity thanks to Behavioral Prediction. In: SAB 2000 From Animals to Animats, the 6th International Conference on the Simulation of Adaptive Behavior, Paris, Meyer, pp. 295–303 (September 2000)

Panzoli, D., Luga, H., Duthen, Y.: Introducing an Associative Memory in a Neural Controller for Advanced Situated Agents. In: Plemenos, D. (ed.) International Conference on Computer Graphics and Artificial Intelligence (3IA 2007), Athens. Greece. Laboratoire XLIM - Université de Limoges, pp. 137–149 (May 2007)

Ray, T.S.: Aesthetically evolved virtual pets. In: Artificial Life 7 workshop proceedings, p. 27, 33, 36, pp. 158–161 (2000)

Sims, K.: Artificial evolution for computer graphics, SIGGRAPH 1991. In: Proceedings of the 18th annual conference on Computer graphics and interactive techniques, pp. 319–328. ACM Press, New York (1991)

Sims, K.: Evolving 3d morphology and behavior by competition. In: Artificial Life IV Proceedings, pp. 28–39. MIT Press, Cambridge (1994)

Sanchez, S.: Solveur de contraintes géométriques pour la modélisation déclarative, Master Recherche de l'école doctorale IIL, Equipe VORTEX-IRIT, Juin (2000)

Sanchez, S., Luga, H., Duthen, Y.: Learning classifier systems and behavioural animation of virtual characters. In: Gratch, J., Young, M., Aylett, R., Ballin, D., Olivier, P. (eds.) IVA 2006. LNCS (LNAI), vol. 4133, p. 467. Springer, Heidelberg (2006)

Sanchez, S., Roux, O., Luga, H., Gaildrat, V.: Constraint-Based 3D Object Layout using a Genetic Algorithm. In: 3IA 2003, The Sixth International Conference on Computer Graphics and Artificial Intelligence, Limoges, 14/05/03-15/05/03 (2003)

Terzopoulos, D., Tu, X.: Artificial fishes: Autonomous locomotion, perception, behavior, and learning in a simulated physical world. Artificial Life 1(4), p. 19, 327–351 (1994)

Todd, S., Latham, W.: Evolutionary Art and Computers, Orlando, FL, USA. Academic Press, Inc., London (1994)

Van De Panne, M.: Sensor-actuator networks. In: SIGGRAPH 1993: Proceedings of the 20th annual conference on Computer graphics and interactive techniques, p. 19, 28, 32, 65, pp. 335–342. ACM, New York (1993)

Ventrella, J.: Explorations in the emergence of morphology and locomotion behavior in animated characters. In: Proceedings of the 4th International Workshop on the Synthesis and Simulation of Living Systems Artificial Life IV, p. 47, pp. 436–441. MIT Press, Cambridge (1994)

Ventrella, J.: Designing emergence in animated artificial life worlds. In: VW 1998: Proceedings of the first International Conference on Virtual Worlds, London, UK, p. 17, 47, pp. 143–155. Springer, Heidelberg (1998)

Whitelaw, M.: Metacreation: Art and Artificial Life. MIT Press, Cambridge (2004)

Wilson, S.W., Zcs,: A zeroth level classifier system. Evolutionary Computation 2(1), 1–18 (1994)

# 8

# User Profiling from Imbalanced Data in a Declarative Scene Modelling Environment

Georgios Bardis[1], Georgios Miaoulis[1], and Dimitri Plemenos[2]

[1] Department of Computer Science, Technological Education Institute of Athens,
Ag.Spyridonos St., 122 10 Egaleo, Greece
gbardis@teiath.gr, gmiaoul@teiath.gr
[2] Laboratoire XLIM, Faculté des Sciences, Université de Limoges
83, rue d'Isle, 87060 Limoges cedex, France
plemenos@unilim.fr

**Abstract.** Declarative Modelling is an early-phase design technique allowing the user to describe an object or an environment in abstract terms, closer to human intuition. The geometric solutions automatically yielded for such a description are evaluated by the user and may be subsequently used for the construction of a computational model of his/her preferences. Due to the physical limitations of the human evaluator, and the large number of the representations produced, only a subset of the latter are actually evaluated by the user and eventually a small number of them are approved, leading to imbalanced datasets in regard to the learning mechanism invoked. In the current work we discuss and assess the capability of a mechanism adopted for user modelling in a declarative design environment to handle this imbalance. The experimental results in this context indicate considerable efficiency in the prediction for the underrepresented class.

**Keywords:** Declarative Modelling, Machine Learning, Multi-criteria Decision Making, Imbalanced Datasets.

## 8.1 Introduction

Declarative Modelling [13] allows abstract description of an object or environment, using declarative terms instead of concrete geometric values and properties. Declarative Modelling by Hierarchical Decomposition [18] enhances this methodology by allowing intermediate objects, grouping the terminal nodes of the description tree with respect to topology, common use, etc. The abstract description is used as the basis for the generation of geometric models that conform to it, yet represent different interpretations of the input. These models are visualised and the user is subsequently able to assess them according to preferences and either terminate the process or refine the description thus restarting the Declarative Modelling cycle. The overall process is typically supported by an environment handling the computationally demanding tasks of solution generation and visual presentation of their internal representation, as shown in Figure 8.1.

D. Plemenos, G. Miaoulis (Eds.): Arti. Intel. Techn. for Comp. Graph., SCI 159, pp. 123–140.
springerlink.com                                    © Springer-Verlag Berlin Heidelberg 2009

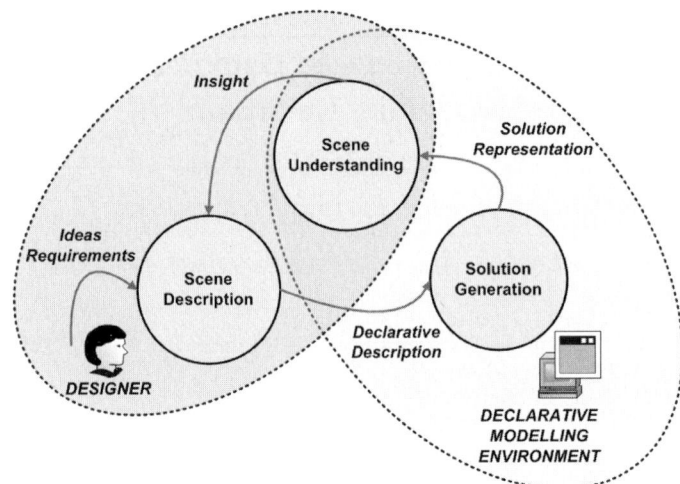

**Fig. 8.1.** Declarative Modelling Cycle

A number of tools have been based on the aforementioned cycle [5],[15],[22], [23],[27]. In regard of the first stage, the construction of the description may involve more than one designer in a collaborative scheme [6]. The stage of geometric model generation may explore the entire solution space, reducing the task to a constraint satisfaction problem [3],[20] or employ genetic techniques [14],[24] thus integrating the second and third stages in a sub-cycle of production and evaluation of subsequent generations. Moreover, selected and possibly modified solutions may be used to reconstruct a declarative description [10]. In any case, solution evaluation may be automated either based on fixed rules [9] or on previous behaviour [1]. In the current work we examine the latter in the context of imbalanced recorded user preferences and application to newly generated solution sets. The employed mechanism aims to acquire, represent and manage each user's preferences as a computational model capable of providing intelligent behaviour features to a declarative design environment. However, in order to achieve this, such a mechanism has to tackle the problem of imbalanced datasets which is inherent in this category of environments.

In the following we initially discuss Declarative Modelling with regard to this imbalance. Next, we present the characteristics of a Declarative Modelling environment and the implied requirements for the corresponding user preference mechanism. Related works are discussed in the subsequent section, where major differences and advantages of the current work are highlighted. Existing techniques for handling imbalance are presented in the next section, being connected with the discussion of the user preference mechanism and its functionality that immediately follows. The setup of the mechanism and the performance metrics used are outlined in the subsequent section. The detailed presentation of the main series of experiments forms the last part of this chapter which is concluded with an overview of the results.

**Table 8.1.** Declarative Description Constituent Primitives

| Constituent | Samples |
|---|---|
| Declarative Relation | near |
| | adjacent-east |
| | wider-than |
| | etc. |
| Declarative Property | long |
| | small |
| | etc. |
| Declarative Object | Kitchen |
| | Bedroom |
| | Storage |
| | Garage |
| | etc. |

## 8.2   Evaluation Imbalance in Declarative Modelling

The main problem that arises in the effort of taking advantage of previous evaluations is the quality of the evaluations when viewed as a training dataset. In particular, a declarative description of only a few declarative objects, interrelated through declarative relations and featuring declarative properties, may lead to hundreds of thousands of solutions. The latter are all valid in terms of compliance with the input description, yet they represent varied responses to each user's preferences. Even in cases where the model is simple, represented by a set of bounding boxes, the user may favour certain solutions due to morphological features not covered by the declarative repertoire of objects, relations and properties. Samples of these declarative description constituents, for simplified building design, appear in Table 8.1.

A simple declarative description of only two objects, one relation and two properties appears in Figure 8.2 whereas example alternative solutions (of the 539 generated for a restricted version of the specific description) appear in Figure 8.3.

In the given subset of solutions a user might approve only solution (ii), for example, due to the coaxial placement of the two volumes. It becomes apparent that even small morphological variations among descendants of the same declarative description may represent the difference between user approval and rejection. Moreover, different users may select different solutions as the most preferable ones even if they have submitted identical declarative descriptions. Therefore, capturing and modelling user preferences implies actual user evaluation thus introducing a number of issues connected with particularities and limitations of the human evaluator.

Based on experimental results from an existing declarative modelling environment it appears to be the case that users tend to approve only a fraction of the solutions

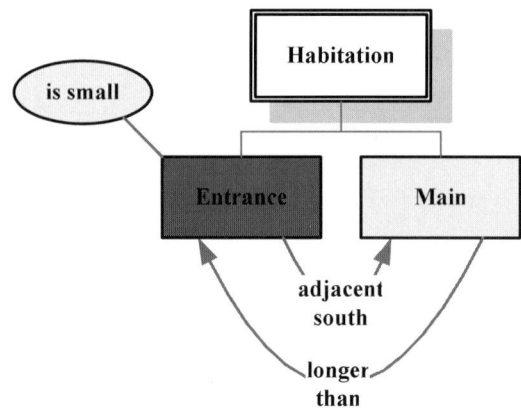

**Fig. 8.2.** Simple Declarative Description

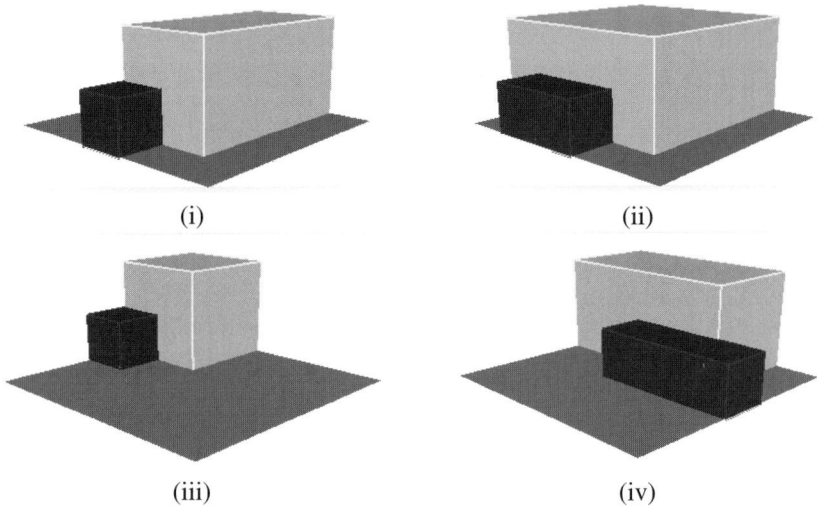

**Fig. 8.3.** Example Alternative Solutions for the Declarative Description of Figure 8.2

they have the opportunity to evaluate. Moreover, it is practically infeasible to evaluate all solutions complying with a given declarative description due to their large numbers and human limitations. In the following, we study the environment characteristics, posing specific requirements for a learning mechanism.

## 8.3  Environment Characteristics

A Declarative Modelling environment imposes a set of restrictions originating from its inherent functionality. Complications also arise from the fact that users may be contradictive in their behaviour depending on context, by approving solutions that would be rejected if inspected among better ones.

We may summarise these restrictions as follows:

- New solutions are produced and evaluated at distinct and usually distant times.
- Storing all produced solutions is prohibitive in terms of space.
- Retraining of mechanism using all produced solutions is prohibitive in terms of time and profile portability.
- Different descriptions may lead to large differences in the size of solution population.
- User may be willing to evaluate only a few or, alternatively, all of the generated solutions.
- Small morphological differences may suggest large differences in the degree of user's preference.
- A declarative description may represent a whole class of solutions the user prefers or, alternatively, may lead to zero or very few approvals.

These characteristics impede the application of traditional machine learning mechanisms and have led to the use of a variation [1] of an incremental learning algorithm [21]. Before we present the basic operation of the algorithm and the techniques employed in the current context, we briefly discuss the works that have attempted to address the problem of acquisition and application of user preferences in the area of Declarative Modelling as well as the existing techniques for handling the class representation imbalance.

## 8.4   Related Works

Although numerous efforts have been presented in the area of geometric design, indicatively [7],[11], in the context of Declarative Modelling there are only a few. In particular, the works that have attempted to capture and utilise user preferences in the Declarative Modelling context are [4] and [19].

In [19] two alternative approaches are followed. According to the first, each user's preferences are modelled by a dedicated neural network which is trained during regular system use by the approved and rejected solutions by the specific user. This approach constructs a separate user profile for each scene that has been submitted for solution generation to the system by the specific user. The alternative approach presented in the same work is based on genetic techniques, using the actual user's evaluation as the fitness function for each generation of solutions. In this manner, generation by generation, the user's preferences are gradually expressed at a higher degree in the solution set. Overall, both approaches presented in [19] operate under the assumption of the *solution search* mode of the Declarative Modelling methodology – where only one or a few, similar and possibly not optimal, solutions are desired by the user – instead of the *exploration mode* – where all solutions that would be approved by the user are desired. The latter is the policy adopted in the current work. Moreover, multiple profiles are created for the same user, one for each submitted description, represented by the dedicated neural networks, whereas the Intelligent User Profile Module creates and maintains a unique user profile for each user, which is applicable to all previous and future descriptions submitted by him/her.

The work presented in [4] focuses on the definition of a target concept via the classification of similar solutions together in classes by an unsupervised classifier, and the subsequent manual evaluation of class representatives on behalf of the user. It also exhibits important differences to the Intelligent User Profile Module in the sense that the notion of target concept is similar to the solution search mode, and, moreover, the unsupervised classifier groups similar solutions together, implying similar user preference for these solutions. The Intelligent User Profile Module, on the other hand, operates in exploration mode and makes no assumption regarding the user preferences even for morphologically similar solutions: the user is free to approve a solution and discard another one, very similar to the first.

## 8.5  Existing Techniques

Due to the collective consideration of error rates, traditional classifiers tend to adapt to the over-represented classes, a fact caused by the error reduction this favouritism implies. Moreover, classes of only a few representatives in the entire set may also be considered as noise. A number of techniques have been applied to handle this imbalance [12],[26],[28] – [25] for an overview.

One family of these proposed techniques operate at the data level applying alternative mechanisms, including:

- Replication of samples from the minority class
- Elimination of samples from the over-represented class.
- Generation of synthetic minority class samples through extrapolation of existing ones.
- Separation of the initial set into balanced subsets where the minority class is replicated but the majority representatives are different in each subset.
  The alternative approach operates at the algorithm level, including:
- Assignment of cost or penalty for misclassification of the minority class samples.
- Application of alternative values for the classification threshold.
- Application of multiple classifiers, typically trained with different subsets, reaching consensus by majority voting.

## 8.6  Applied Mechanism

The declarative modelling cycle starts with the construction of a new declarative description or the alteration of an existing one. In any case, the description is submitted to the solution generator which, in turn, yields geometric models of the scene, all conforming to the description.

### Solution Encoding

In order to apply machine learning techniques on the geometric models, the latter are encoded as vectors of values with respect to a pre-defined set of observed

**Table 8.2.** Indicative Observed Attributes for Residence Project Type

| Observed Attribute | Range |
| --- | --- |
| Private Zone Area | `0..maxprivatezone` |
| Public Zone Area | `0..maxpubliczone` |
| Non-oblong Rooms Percentage | `0..100` |
| Private/Public Zone Separation | `false, true` |
| Southwestern Bedroom | `false, true` |

morphological characteristics, directly extracted from the geometric representation. These vectors coupled with the user's evaluation are used for all training and evaluation purposes of the mechanism.

The observed morphological attributes and their parameters vary for different kinds of buildings, i.e. different project types. Table 8.2 presents an indicative set of observed attributes for the Residence project type. Only residential spaces representing terminal nodes of the part of hierarchical decomposition of the declarative description are considered at this stage.

Each attribute is calculated from the elementary geometric properties of each space and represents the performance of the specific solution against the specific attribute. For example, the geometric interpretation of the Private/Public zone separation attribute is connected to the validity of the logical statement:

$$\max(x_{opr}+l_{opr}) \leq \min(x_{opu}) \vee \max(y_{opr}+w_{opr}) \leq \min(y_{opu}) \vee$$
$$\max(x_{opu}+l_{opu}) \leq \min(x_{opr}) \vee \max(y_{opu}+w_{opu}) \leq \min(y_{opr})$$

where *opr* and *opu* represent any object of the private or the public zone respectively, whereas *x,y,l,w* represent the x coordinate, the y coordinate, the length and the width of such an object. The intuitive interpretation of this observed attribute is explicitly shown by the environment in the example session of Figure 8.4. where its value is *false*.

## Operation

Due to the requirement for conformity with the functionality of the Declarative Modelling environment summarised in the previous section, traditional learning mechanisms were not applicable in the current context. The employed mechanism consists of a committee of neural networks, that is expanded with a new sub-committee, i.e. new members, every time a new scene has been submitted and the produced solutions have been manually evaluated by the user. The algorithm for the committee training and expansion is based on Learn++ [21] which in turn belongs to the family of AdaBoost algorithms [8].

The rationale of this category of algorithms is the combination of many weak learners to form a committee of strong learner behaviour. In the current approach, samples are selected for training randomly but using their normalised weights as probabilities. Initially, all weights are equal. A new learner is added to the committee after training, as long as it does not fall under the weak learner error threshold of 0.5

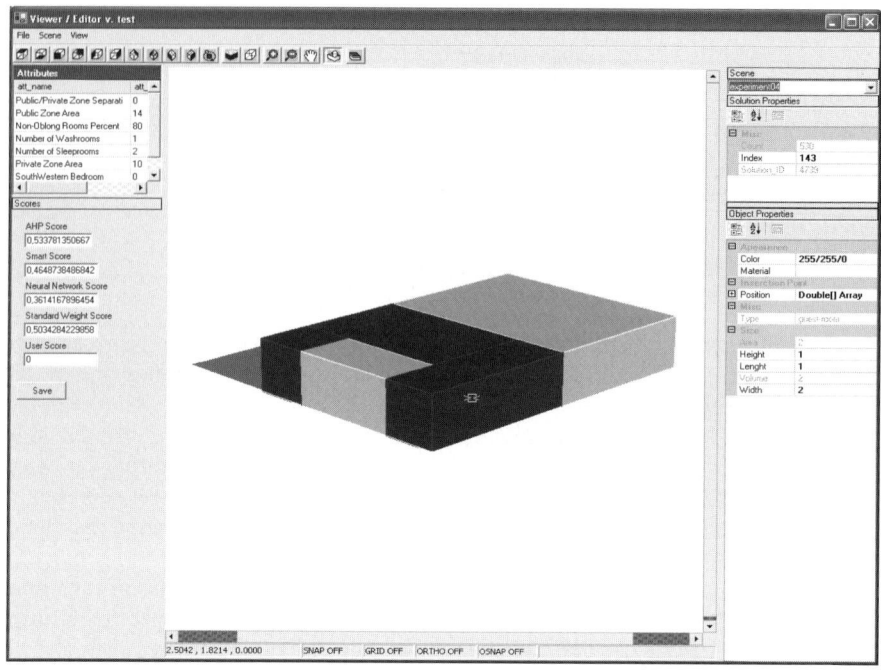

**Fig. 8.4.** Example Session: Private/Public Zone Separation is *false* for Visualised Solution

while the overall error of the committee, including the new learner, is also maintained above the same threshold.

Once a new member is added to the committee, the remaining samples are re-evaluated and re-weighted according to overall classification: misclassified samples retain their weights whereas correctly classified samples have theirs reduced. In this way, misclassified samples have increased probability to be included in the next training set thus increasing the probability of the creation of a new learner correctly classifying them. The variation of the algorithm used in the current work stops including samples to the new training set as soon as the sum of their weights surpasses half the total sum of weights, whereas the original algorithm continued to expand the new training set until half of the count of available samples were included.

## 8.7  Discussion

The policy applied to the inclusion of the new weak learner allows concentration of the latter on the hard-to-classify samples instead of being trained with a lot of easy examples and a few difficult ones in order to reach the half of the total number of available samples. This change has accelerated the generation of adequate weak learners especially in advanced stages of the training process [1]. In other words, this modification has allowed the creation of weak learners solely or largely trained by minority samples due to their increased weight. Moreover, the formation of the new subcommittee stops as soon as its error falls below a threshold of adequacy. In this way, sub-committees are adjusted to the difficulty of the current dataset.

In connection with the policies for handling imbalance, the mechanism practically implements the algorithmic approaches of cost assignment to misclassified minority representatives. This is achieved through the aforementioned weighting policy. Moreover, since the overall evaluation of any sample is the result of a majority voting, which takes into account the error of each individual weak learner during its inclusion to the committee, the mechanism also applies the approach of multiple classifiers trained with alternative datasets.

## 8.8  Performance Metrics

Thorough evaluation of the mechanism requires a collection of metrics capturing different aspects of its performance. These metrics have been formally defined in [1] and cover the entire range of the mechanism behaviour. We present a brief definition of these indices below:
Let

A the set of all generated solutions for a specific description.

G the set of all user evaluated solutions for a specific description.

U the set of evaluated solutions approved by the actual user for a specific description. By definition $U \subseteq G$.

M the set of evaluated solutions approved by the mechanism for a specific description. By definition $M \subseteq G$.

It follows from the definitions above that:

M-U the set of evaluated solutions approved erroneously by the mechanism.

U-M the set of evaluated solutions rejected erroneously by the mechanism.

The latter are visually demonstrated in Figure 7.5.

The following indices have been used for the experimental evaluation:

- Error Ratio: ER= $\dfrac{|M-U|+|U-M|}{|G|}$

- Hit Ratio: HR= $\dfrac{|M \cap U|}{|U|}$

- Performance Ratio: PR= $\dfrac{|M \cap U|}{|M|}$

Each one of these ratios captures a different performance aspect. In particular, ER presents the overall percentage of generated solutions that were erroneously classified by the mechanism. HR signifies the percentage of user approved solutions that the mechanism discovered, thus measuring the mechanism's ability to recall. PR is the percentage of the mechanism approved solutions that are actually approved by the user, indicating the mechanism's precision. Notice that some extreme cases may yield

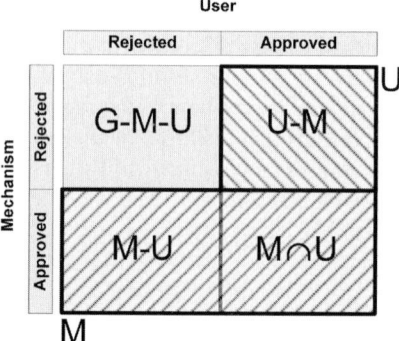

**Fig. 8.5.** Visual Explanation of U, M, G sets and their Interconnection

high values for an index without necessarily representing an acceptable approach. As an example we may consider a mechanism that blindly approves all generated solutions and, thus, approves all solutions also approved by the user. This mechanism would yield a high HR (high recall), however, its failure to accurately model the user's preferences would be revealed by an extremely high ER and low PR (low precision). On the other extreme, a mechanism approving only one solution which is also approved by the user would yield a high PR (high precision), but high ER and low HR (low recall).

## 8.9   Dataset and Mechanism Configurations

A number of experiments have been performed using the aforementioned mechanism, under varying values of its parameters. It came as no surprise that configurations exhibiting the lowest overall error rates were not equally performing in terms of minority class prediction. Five alternative abstract scene descriptions have been used for the gradual training of the mechanism with evaluated solutions from each scene separately. This policy was adopted to simulate the regular functionality of the environment with respect to user sessions.

Table 8.3 summarises the proportions of the evaluated samples with respect to the overall generated populations.

**Table 8.3.** Overall and Evaluated Solution Populations

| Declarative Scene Description | Overall Solution Population | Evaluated Solutions | % |
|---|---|---|---|
| Naxos Habitation | 670260 | 670 | 0.10% |
| Rentis Habitation | 21184 | 411 | 1.94% |
| Piraeus Habitation | 306255 | 621 | 0.20% |
| Kalamaki Habitation | 212096 | 530 | 0.25% |
| Patras Habitation | 511450 | 511 | 0.10% |

**Table 8.4.** Inspected against Approved Solutions for Different Users

| Declarative Scene Description | Inspected Solutions | Approved Solutions (Sophia) | Approved Solutions (Katerina) |
|---|---|---|---|
| Naxos  Habitation | 670 | 41 | 130 |
| Rentis Habitation | 411 | 33 | 125 |
| Piraeus Habitation | 621 | 47 | 114 |
| Kalamaki Habitation | 530 | 17 | 42 |
| Patras Habitation | 511 | 52 | 50 |

**Table 8.5.** Configurations of Learning Mechanism

| Parameter | Conf.1 | Conf.2 | Conf.3 |
|---|---|---|---|
| weak learners added per scene (WL) | 5 | 10 | 10 |
| first hidden layer neurons (HL1) | 4 | 4 | 4 |
| second hidden layer neurons (HL2) | 0 | 0 | 0 |
| epochs (E) | 1000 | 1000 | 1000 |
| learning rate (LR) | 0.028 | 0.02 | 0.028 |
| momentum (M) | 0.5 | 0.5 | 0.5 |
| error goal (EG) | 0.1 | 0.1 | 0.1 |

It is worth noting that even for these reduced numbers – at least compared with the overall populations – users were usually hesitant both in evaluating the entire subset manually as well as in evaluating solutions from more than one or two descriptions. These facts reveal the inherent difficulties of a functional declarative modelling environment and underline the need for efficient user profiling.

Table 8.4 summarises the ratio of approved solutions against the count of solutions actually inspected by a user. The under-representation of solutions successfully expressing user preferences becomes evident here. Moreover, it can be observed that the mere number of available solutions does not necessarily imply a proportional number of approvals on behalf of the user. This case appears in the fourth row of Table 8.4 where an amply populated scene receives only a few approvals from the first user.

Three different configurations of the learning mechanism have been employed in the series of experiments presented in the next section. They are the best performing representatives of numerous alternative configurations that have been checked against a variety of datasets and metrics.

## 8.10   Experimental Results

Experiments were conducted upon gradual training, forming training sets of solutions from only one scene at every stage. After each stage of training, the generalisation ability of the mechanism was evaluated against previously unseen scenes, i.e. solutions originating from descriptions that had contributed not even a single solution to the training of the mechanism. This approach has been employed in order to simulate regular system use of a declarative modelling environment.

### Overall Generalisation after First Scene

Table 8.6 demonstrates the differences of three alternative configurations in terms of performance with respect to overall generalisation error after the mechanism has been trained with evaluated solutions from only one scene (Naxos Habitation).

Configurations refer to internal calibration of the neural networks playing the role of weak learners as well as to the overall committee settings. Training has been based on evaluations from a single user and it is interesting to observe that although there is clear dominance of the second configuration in overall performance, this superiority is not reflected to every scene.

**Table 8.6.** Overall Generalisation Error (Dataset from user Sophia)

| Generalisation ER | Conf. 1 | Conf. 2 | Conf. 3 |
|---|---|---|---|
| Rentis Habitation | 27,74% | 34,31% | 45,50% |
| Piraeus Habitation | 67,47% | 24,80% | 81,16% |
| Kalamaki Habitation | 25,09% | 13,21% | 35,66% |
| Patras Habitation | 10,18% | 22,90% | 37,18% |
| Overall | 34,64% | 23,25% | 51,62% |

**Table 8.7.** Overall Generalisation Error (Dataset from user Katerina)

| Generalisation ER | Conf. 1 | Conf. 2 | Conf. 3 |
|---|---|---|---|
| Rentis Habitation | 12,90% | 12,41% | 19,71% |
| Piraeus Habitation | 10,14% | 19,00% | 25,60% |
| Kalamaki Habitation | 3,40% | 5,85% | 6,98% |
| Patras Habitation | 14,09% | 22,31% | 18,00% |
| Overall | 9,94% | 15,15% | 17,80% |

Results from another user's evaluation appear in Table 8.7. It is interesting to observe that the previously superior second configuration is now superseded by the first, although it exhibits improved performance. The overall error for all configurations is significantly decreased, mainly revealing consistency in the selections of the specific user that allow the mechanism to safely generalise.

### Minority Class Generalisation after First Scene

In Tables 8.8 and 8.9, we focus the generalisation ability of the mechanism on the under-represented class of user-approved solutions. We may observe the superiority of Configuration 3 against the other two for both users. We may also notice important differences in performance of each configuration depending on the scene.

Next, we explore the mechanism's precision by examining the values of its performance ratio during the same experiments. It is interesting to observe, in Tables 8.10, 8.11 that Configuration 3 is not the dominant one any more. The overall metrics reveal that

**Table 8.8.** Minority Class Generalisation Error (Dataset from user Sophia)

| Minority Class Generalisation – HR | Conf. 1 | Conf. 2 | Conf. 3 |
|---|---|---|---|
| Rentis Habitation | 54,40% | 68,80% | 90,40% |
| Piraeus Habitation | 97,37% | 90,35% | 100,00% |
| Kalamaki Habitation | 95,24% | 2,38% | 95,24% |
| Patras Habitation | 30,00% | 78,00% | 94,00% |
| Overall | 70,69% | 69,18% | 94,86% |

**Table 8.9.** Minority Class Generalisation Error (Dataset from user Katerina)

| Minority Class Generalisation – HR | Conf. 1 | Conf. 2 | Conf. 3 |
|---|---|---|---|
| Rentis Habitation | 45,45% | 57,58% | 63,64% |
| Piraeus Habitation | 74,47% | 78,72% | 89,36% |
| Kalamaki Habitation | 0,00% | 5,88% | 5,88% |
| Patras Habitation | 69,23% | 86,54% | 84,62% |
| Overall | 57,72% | 68,46% | 72,48% |

**Table 8.10.** Minority Class Generalisation Error (Dataset from user Sophia)

| Minority Class Generalisation – PR | Conf. 1 | Conf. 2 | Conf. 3 |
|---|---|---|---|
| Rentis Habitation | 54,40% | 45,74% | 39,24% |
| Piraeus Habitation | 21,06% | 41,87% | 18,45% |
| Kalamaki Habitation | 23,39% | 3,33% | 17,62% |
| Patras Habitation | 46,88% | 26,90% | 20,09% |
| Overall | 27,37% | 37,60% | 22,97% |

**Table 8.11.** Minority Class Generalisation Error (Dataset from user Katerina)

| Minority Class Generalisation – PR | Conf. 1 | Conf. 2 | Conf. 3 |
|---|---|---|---|
| Rentis Habitation | 30,00% | 33,93% | 23,33% |
| Piraeus Habitation | 40,70% | 25,52% | 21,43% |
| Kalamaki Habitation | 0,00% | 6,25% | 4,55% |
| Patras Habitation | 39,13% | 29,61% | 34,38% |
| Overall | 37,55% | 27,64% | 24,77% |

the mechanism is considerably more successful in recall than in precision. Intuitively, the mechanism misinterprets a number of rejected solutions as approved; however, it manages to pinpoint the major part of solutions the user has approved.

**Generalisation after Second Scene**

Tables 8.12, 8.13 and 8.14 present the evolution of both the overall generalisation and the minority generalisation errors for the same user (previous values appear in parentheses). The mechanism has processed evaluated solutions from an additional scene for training and subsequently has been assessed against the unseen samples from the remaining scenes.

**Table 8.12.** Overall Generalisation Error (Dataset from user Sophia)

| Generalisation Error | Conf. 1 | | Conf. 2 | | Conf. 3 | |
|---|---|---|---|---|---|---|
| Piraeus Habitation | 31,72% | (67,47%) | 41,06% | (24,80%) | 33,01% | (81,16%) |
| Kalamaki Habitation | 8,30% | (25,09%) | 14,72% | (13,21%) | 9,81% | (35,66%) |
| Patras Habitation | 36,79% | (10,18%) | 43,05% | (22,90%) | 23,29% | (37,18%) |
| Overall | 25,81% | (34,64%) | 33,27% | (23,25%) | 22,62% | (51,62%) |

**Table 8.13.** Minority Class Generalisation Error (Dataset from user Sophia)

| Minority Class Gen.- HR | Conf. 1 | | Conf. 2 | | Conf. 3 | |
|---|---|---|---|---|---|---|
| Piraeus Habitation | 98,25% | (97,37%) | 94,74% | (90,35%) | 100,00% | (100,00%) |
| Kalamaki Habitation | 47,62% | (95,24%) | 57,14% | (2,38%) | 61,90% | (95,24%) |
| Patras Habitation | 100,00% | (30,00%) | 100,00% | (78,00%) | 88,00% | (94,00%) |
| Overall | 88,35% | (70,69%) | 88,35% | (69,18%) | 89,32% | (94,86%) |

The overall generalisation error is reduced for the first and considerably reduced for the third configuration and increased for the second which is no longer the best performing one in terms of this metric. Each additional dataset used for training, originating from a single declarative description, potentially introduces additional aspects of the user's preferences. Hence, the performance of the mechanism may deteriorate in trying to capture these aspects while maintaining previous knowledge. Notice that the mechanism is incrementally trained using only samples from the current dataset whereas previous knowledge is maintained solely through the existing members of the committee.

All configurations exhibit varied performance in terms of the minority class of user-approved solutions, depending on the evaluated scene. Once again, the effort to acquire additional knowledge inevitably leads to partial performance degradation as is the case with the Kalamaki scene, where the first and third configurations demonstrate reduced performance. On the other hand, the first configuration exhibits a steep increase in performance with respect to the Patras scene. It is also interesting to observe that even for scenes where the minority representation falls below 10%, as shown in Table 8.4, the mechanism achieves adequate performance as in the case of Patras habitation. Nevertheless, the introduction of additional samples has a positive

**Table 8.14.** Minority Class Generalisation Error (Dataset from user Sophia)

| Minority Class Gen. - PR | Conf. 1 | | Conf. 2 | | Conf. 3 | |
|---|---|---|---|---|---|---|
| Piraeus Habitation | 36,48% | (21,06%) | 30,25% | (41,87%) | 35,74% | (18,45%) |
| Kalamaki Habitation | 47,62% | (23,39%) | 28,57% | (3,33%) | 41,94% | (17,62%) |
| Patras Habitation | 21,01% | (46,88%) | 18,52% | (26,90%) | 28,03% | (20,09%) |
| Overall | 31,01% | (27,37%) | 25,60% | (37,60%) | 34,20% | (22,97%) |

**Table 8.15.** Configuration Performance Summary (Dataset from user Sophia)

| Metrics | Experimental Stage | Conf. 1 | Conf. 2 | Conf. 3 |
|---|---|---|---|---|
| Overall Error | After 1st scene | 34,64% | 23,25% | 51,62% |
| | After 2nd scene | 25,81% | 33,27% | 22,62% |
| Overall HR | After 1st scene | 70,69% | 69,18% | 94,86% |
| | After 2nd scene | 88,35% | 88,35% | 89,32% |
| Overall PR | After 1st scene | 27,37% | 37,60% | 22,97% |
| | After 2nd scene | 31,01% | 25,60% | 34,20% |

effect to the generalisation ability with respect to the minority class. In particular, the first and second configuration achieve hit rates closer to 0.9 whereas the third configuration has reduced performance but still maintains a similar hit ratio. In regard to performance ratio, Configurations 1 and 3 exhibit moderate improvement.

This fact, coupled with the improved overall generalisation performance of these configurations reveal the capability of the mechanism to incrementally learn user preferences through a series of evaluated solutions.

Table 8.15 summarises the evolution of error and ratios for each configuration where it becomes apparent that no single configuration performs better than the others in every case.

## 8.11 Conclusions

The attempt to create a computational model of user preferences in a declarative modelling environment raises a number of issues due to restrictions inherent to the functionality of such an environment. These issues partly originate from the inability of the human user to evaluate large numbers of solutions that could be used for training the mechanism and from the fact that, even when evaluations are submitted, they usually contain only a small fraction of approved samples. The latter leads to the problem of imbalanced class representation that has to be addressed in the special context of declarative modelling.

In the current paper we have examined the performance of an approach for user preference acquisition under the light of known techniques for handling imbalanced datasets. The mechanism employed learns user preferences incrementally, being augmented at each time a new dataset is available and adopts techniques for imbalance resolution at the algorithm level.

The experiments reveal that the mechanism employed has the ability to acquire and enhance its knowledge of user's preferences. It performs at satisfactory levels in the majority of the cases, especially with respect to its recall ability. Nevertheless, it has also become apparent that performance may vary depending on user's preference consistency or contradictions and the corresponding data subsets, revealed by the fact that no single configuration had consistently better performance throughout the experiments.

## References

1. Bardis, G.: Machine Learning and Decision Support for Declarative Scene Modelling / Apprentissage et aide à la décision pour la modélisation déclarative de scènes (bilingual), Thèse de Doctorat, Université de Limoges, France (2006)
2. Bardis, G., Golfinopoulos, V., Makris, D., Miaoulis, G., Plemenos, D.: Experimental Results of Selective Visualisation According to User Preferences in a Declarative Modelling Environment. In: 10th 3IA – International Conference on Computer Graphics and Artificial Intelligence Infographie Interactive et Intelligence Artificielle, Athens, Greece, pp. 29–38 (2007) ISBN 2-914256-09-4
3. Bonnefoi, P.-F., Plemenos, D.: Constraint Satisfaction Techniques for Declarative Scene Modelling by Hierarchical Decomposition. In: 4th 3IA – International Conference on Computer Graphics and Artificial Intelligence, Limoges, France, pp. 89–102 (2002) ISBN 2-914256-03-5

4. Champciaux, L.: Classification: A Basis for Understanding Tools in Declarative Model-ling. Computer Networks and ISDN Systems 30, 1841–1852 (1998)
5. Chauvat, D.: The VoluFormes Project: An Example of Declarative Modelling with Spatial Control, PhD Thesis, Nantes, France (1994)
6. Dragonas, J.: Collaborative Declarative Modelling / Modelisation Declarative Collabora-tive (bilingual), Thèse de Doctorat, Université de Limoges, France (2006)
7. Essert-Villard, C., Schreck, P., Dufourd, J.-F.: Sketch-based pruning of a solution space within a formal geometric constraint solver. Artificial Intelligence 124, 139–159 (2000)
8. Freund, Y., Schapire, R.: A decision-theoretic generalization of online learning and an ap-plication to boosting. Journal of Computer and System Sciences 55(1), 119–139 (1997)
9. Fribault, P.: Modelisation Declarative d'Espaces Habitable (in French), Thèse de Doctorat, Université de Limoges, France (2003)
10. Golfinopoulos, V.: Study and Implementation of a Knowledge-based Reverse Engineering System for Declarative Scene Modelling / Étude et réalisation dun système de rétro-conception basé sur la connaissance pour la modélisation déclarative de scènes (bilingual), Thèse de Doctorat, Université de Limoges, France (2006)
11. Joan-Arinyo, R., Luzon, M.V., Soto, A.: Genetic algorithms for root multi-selection in constructive geometric constraint solving. Computers and Graphics 27, 51–60 (2003)
12. Kotsiantis, S., Tzelepis, D., Koumanakos, E., Tampakas, V.: Selective Costing Voting for Bankruptcy Prediction. International Journal of Knowledge-Based & Intelligent Engineer-ing Systems (KES) 11(2), 115–127 (2007)
13. Lucas, M., Martin, D., Martin, P., Plemenos, D.: The ExploFormes project: Some Steps Towards Declarative Modelling of Forms. In: AFCET-GROPLAN Conference, Strasbourg (France), November 29 – December 1, vol. 67, pp. 35–49. Published in BIGRE (1989) (in French)
14. Makris, D.: Study and Realisation of a Declarative System for Modelling and Generation of Style with Genetic Algorithms: Application in Architectural Design / Etude et réalisa-tion d'un système déclaratif de modélisation et de génération de styles par algorithmes gé-nétiques: application à la création architecturale (bilingual), Thèse de Doctorat, Université de Limoges, France (2005)
15. Martin, D., Martin, P.: PolyFormes: Software for the Declarative Modelling of Polyhedra. The Visual Computer 15, 55–76 (1999)
16. Miaoulis, G.: Contribution à l'étude des Systèmes d'Information Multimédia et Intelligent dédiés à la Conception Déclarative Assistée par l'Ordinateur – Le projet MultiCAD, Thèse de Doctorat, Université de Limoges, France (2006)
17. Miaoulis, G., Plemenos, D., Skourlas, C.: MultiCAD Database: Toward a unified data and knowledge representation for database scene modelling. In: 3rd 3IA International Confer-ence on Computer Graphics and Artificial Intelligence, Limoges, France (2000)
18. Plemenos, D.: Declarative modelling by hierarchical decomposition. The actual state of the MultiFormes project. In: Communication in International Conference GraphiCon 1995, St Petersburg, Russia (1995)
19. Plemenos, D., Miaoulis, G., Vassilas, N.: Machine learning for a General Purpose Declara-tive Scene Modeller. In: International Conference GraphiCon 2002, Nizhny Novgorod, Russia (2002)
20. Plemenos, D., Tamine, K.: Increasing the efficiency of declarative modelling. Constraint evaluation for the hierarchical decomposition approach. In: International Conference WSCG 1997, Plzen, Czech Republic (1997)

21. Polikar, R., Byorick, J., Krause, S., Marino, A., Moreton, M.: Learn++: A Classifier Independent Incremental Learning Algorithm. In: Proceedings of Int. Joint Conf. Neural Networks, pp. 1742–1747 (2002)
22. Poulet, F.: Modélisation déclarative de scènes tridimensionnelles: Le projet SpatioFormes, Infographie Interactive et Intelligence Artificielle (3IA), Limoges (1994)
23. Poulet, F., Lucas, M.: Modelling Megalithic Sites. EuroGraphics 15(3), 279–288 (1996)
24. Vassilas, N., Miaoulis, G., Chronopoulos, D., Konstantinidis, E., Ravani, I., Makris, D., Plemenos, D.: MultiCAD-GA: A System for the Design of 3D Forms Based on Genetic Algorithms and Human Evaluation. In: Vlahavas, I.P., Spyropoulos, C.D. (eds.) SETN 2002. LNCS (LNAI), vol. 2308, pp. 203–214. Springer, Heidelberg (2002)
25. Visa, S., Ralescu, A.: Issues in Mining Imbalanced Data Sets - A Review Paper. In: Proceedings of the Sixteen Midwest Artificial Intelligence and Cognitive Science Conference, MAICS, pp. 67–73, Dayton, April 16-17 (2005)
26. Weiss, G.M., Provost, F.: Learning When Training Data are Costly: The Effect of Class Distribution on Tree Induction. Journal of Artificial Intelligence Research 19, 315–354 (2003)
27. Xu, K., Stewart, J., Fiume, E.: Constraint-based Automatic Placement for Scene Composition, Graphics Interface, Canada (2002)
28. Zhang, J., Mani, I.: k-nn Approach to Unbalanced Data Distributions: A Case Study Involving Information Extraction. In: Proceedings of the ICML-2003 Workshop: Learning with Imbalanced Data Sets II, pp. 42–48 (2003)

# 9

# Collaborative Evaluation Using Multiple Clusters in a Declarative Design Environment

Nikolaos Doulamis[1], George Bardis[1], John Dragonas[1], Georgios Miaoulis[1], and Dimitri Plemenos[2]

[1] Technological Education Institute of Athens Department of Informatics Ag.Spyridonos St., 122 10 Egaleo, Greece
Tel.: (+30) 2 10 53 85 312; Fax: (+30) 2 10 59 10 975
ndoulam@cs.tua.gr
[2] Laboratoire XLIM Faculté des Sciences, Université de Limoges 83, rue d'Isle, 87060 Limoges cedex, France
Tel.: (+33) 5 55 43 69 74; Fax: (+33) 5 55 43 69 77
plemenos@unilim.fr

**Abstract.** Collaborative Declarative Modeling combines the flexibility of abstract declarative description of a scene with the integration of views and opinions of multiple designers. One of the stages where this integration applies is the scene understanding phase, where solutions corresponding to the submitted description are visualized and evaluated. In the current work we propose a mechanism for collaborative solution evaluation based on previous feedback with respect to each collaborating designer's preferences. Designers are grouped into clusters of preference revealing the proximity of their evaluations for previously seen solutions. These clusters are formed upon criteria ensuring wide inter-cluster preference representation and thorough intra-cluster compactness. Clusters subsequently participate through representative designers to the formation of the consensus solution set corresponding to the submitted description. The proposed mechanism allows adjustment of parameters controlling intuitive characteristics of the collaborative declarative design methodology such as importance of each designer and strictness in solution evaluation.

## 9.1 Introduction

The increased processing and communication speeds widely available today allow the implementation of environments and solutions that would otherwise be infeasible. The ability to produce large amounts of valid solutions based on an abstract scene description that is inherent in the Declarative Modelling methodology [7] is now enhanced, coupled with the ability to communicate the description and/or the solutions to multiple, geographically distant, locations in a fraction of the previously needed time. This kind of infrastructure has called for policies offering collaborative schemes [1], [2], [3], [4], [5], [6], [14], [15] that will take advantage of it and allow the integration of the opinions of multiple designers on a multitude of solutions and/or descriptions [11].

Application scenarios of such policies are indeed a lot. Examples include remote designs on architectural or structural data, computer-aided design techniques, development

D. Plemenos, G. Miaoulis (Eds.): Arti. Intel. Techn. for Comp. Graph., SCI 159, pp. 141–157.
springerlink.com

of collaborative learning strategies, etc. Other application domains in which automatic user's modelling would be graceful is in content providers sites so that the retrievals are actually tailored to the final user's information needs and preferences. In these cases, different results will obtained for different scenarios. For instance, in an e-commerce application, different types of products will be displayed to users of common preferences.

It remains the case, however, that the successful merge of views requires efficient processing of previous activity and existing feedback. In the current work we present an approach for collaborative solution selection in the context of a declarative design environment, based on the [16] approach yet adding considerable refinement to the profile estimation and integration mechanisms. The next section includes a presentation of the Declarative Modelling (DM) methodology and highlights the phases where the current work intervenes with respect to the collaborative interpretation of DM. The following section provides an intuitive overview of the proposed mechanism and its comprising modules. These modules are formally defined and explained in detail in the subsequent two sections. Finally, the last section provides the conclusions of the proposed mechanism in the context of collaborative DM.

## 9.2   Collaborative Declarative Modelling

The Declarative Modelling methodology offers a powerful tool for early-phase design. It allows the designer to describe an object or an environment in abstract terms instead of concrete geometric values and properties.

### 9.2.1   DM and DMHD

A typical declarative description consists of a set of declarative objects, their properties and the relations among them. The Declarative Modelling by Hierarchical Decomposition [9], [13] enhances this capability by allowing intermediate, abstract objects, to be used to further refine the initial description. DMHD implies in depth knowledge of the structure – but not the exact form – of the object or the environment to be created and is employed when the complexity of the latter prohibits its decomposition as a single level hierarchy. The declarative description according to DMHD is often represented by a directed acyclic graph which is obtained by the *part-of* decomposition tree enriched by edges revealing the declarative relations among nodes.

Both DM and DMHD methodologies include three generic stages as presented in Fig. 9.1; the *declarative scene description*, the *solution generation* and the *scene understanding* stage.

In the first stage, we describe the environment of the scene we want to design. Scene description is performed in an abstractive, intuitive way. This description is then provided as input to the next stage which is responsible for the solution generation, i.e. the production of geometric models that respect the description's requirements. The solutions are submitted to the next and final stage of scene understanding, during which the geometric models are visualized and presented to the designer who has the chance to obtain insight with respect to the initial description and its implications and, thus, refine it, potentially starting a new cycle.

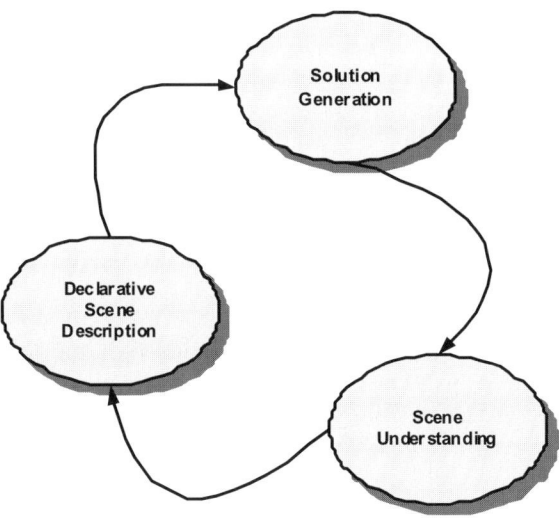

**Fig. 9.1.** The three phases of a declarative modeling environment

However, since a scene is developed under an intuitive manner, the description is often imprecise. The lack of precision is mainly due to two reasons, a) designers do not know the exact properties of the scene (e.g., width, height, exact position of the objects) and b) the ambiguous description of the designer's requirements (e.g., put object A on the left of the object B without determining the exact position on the left). For this reason, the generated solutions at the second stage of the declarative modeling environment (see Fig. 9.1) are too many. This means that filtering mechanisms are required to filter out the generated solutions with respect to the user's information needs.

One way to filter out the generated solutions is described in the [10]. In this framework, the users are allowed to interact with the system by selecting a set of relevant / irrelevant retrieved solutions. These solutions are then *feed back* to the system to adjust its response so as to satisfy the user's information needs. Therefore, [10] implements a designer centric personalization mechanism.

Collaborative Design may interfere with more than one stages of the Declarative Modelling methodology. In particular, designers may collaborate during the scene definition stage [8], [11] incorporating objects and changes yielding a single declarative description. Moreover, collaboration may take place during scene understanding where the designers may provide custom evaluation for alternative solutions, eventually yielding a single solution or a restricted subset representing the overall designers' consensus. For this reason, a collaborative method for estimating the user's preferences in presented in [16]. In particular, in [16] the degree of relevance of a solution is estimated according to average degree of importance by all the users perhaps constrained with the user's significance in a design. It is clear that this scheme provides better results compared to a single user –based adaptation mechanism since more information is exploited.

The main drawback of this approach is that *all the users are handled in a same framework.* However, human perceives content in a subjective rather than in an objective way. Different users have different preferences and interpret the same scene description in a different way. For this reason, taking into consideration the "average" degree of relevance of a solution from all users, we may result into inadequate filtering especially in cases of users of quite different preferences. To address this difficult clustering methods are required in order to classify user's characteristics into different classes of preferences. Then, the consensus process is implemented in the direction of a class preference.

## 9.3   Collaborative Evaluation Using Multiple Clusters

In order to achieve consensus with respect to a given set of solutions, corresponding to a newly submitted description, we take advantage of previously acquired information during the evaluation of past sets of solutions 0, 0. Fig. 9.2 presents the architecture of the proposed system.

As we observed in this figure, the *intelligent user profile module* is responsible for filtering the generated solutions, provided by the *Solution Generator* module. This module is responsible for estimating how close a solution to the preferences of a user

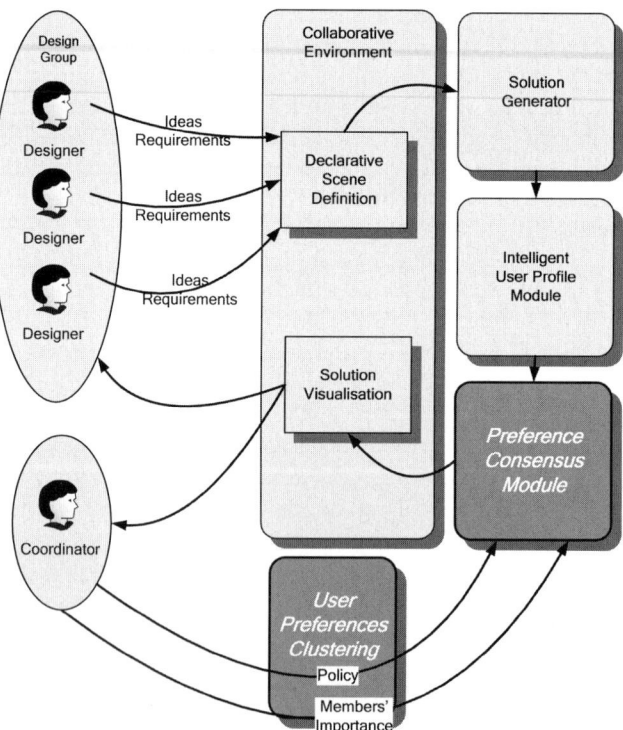

**Fig. 9.2.** The proposed architecture

is. The preference consensus module is responsible for estimating the user's preference by taking into consideration the evaluation from a set of collaborative users instead from a single user itself. This module has been presented in details in the paper of [16]. Instead, in this paper, we extend the architecture of [16] by incorporating a user' preference clustering mechanism. The mechanism is responsible for grouping users with the same subjective characteristics together so as to improve filtering performance.

In particular, each time a user evaluates the retrieved results a set of evaluation descriptors is created. These evaluation descriptors are used for grouping the designers in *clusters of relevance*. These clusters, at the expense of pre-processing, offer the benefit of reduced processing during actual solution evaluation, thus making the procedure of reaching overall consensus more time-efficient.

In particular, each cluster represents a group of designers with similar preferences. The clustering is based on previous evaluations of solutions and its aim is twofold. On the one hand, it attempts to maximise the *density* of each cluster, ensuring it represents a group of close neighbours in the solution evaluation space. On the other hand, it aims to maximise the *preference distance* between different clusters, thus justifying the necessity for the existence of each separate cluster.

Once the clusters have been constructed, a *representative designer* is extracted from each, based on its internal position with respect to the other designers belonging to the same cluster. The aim at this stage is to elicit an individual who's preferences fall, intuitively, *nearest to the centre* of its cluster. These representatives are the ones who will actually *vote* towards the final evaluation of the new set of solutions. This voting may be performed either by the individuals themselves or automatically, upon consideration of their previous evaluations.

Each representative's vote is assigned a custom weight before they are all combined in order to extract the final verdict for each solution, according to the second "weighted" methodology presented in [16]. The evaluation for each solution, on behalf of the representative designers in the current work, is amplified according to the weight assigned to each designer. For example, the overall evaluation for each solution is practically the sum of the weights of all designers approving it in case of binary classification. The weight for each designer in [16] depended on custom factors (seniority, time spent on the description, etc.). In the current work, we extent the weight calculation to include a cluster related metric, associated with the density of the cluster which the designer represents. The aim is to integrate all useful feedback on the specific designer in a uniform index.

## 9.4  Optimal Designers Profile

As stated above, the declarative design is accomplished in three different phases. In the first phase, we declare the description of the scene. In the second phase, the system retrieves relevant solutions with respect to the scene description (phase one). Finally, in the third phase, the designers evaluate the retrieved results according to their information needs.

Let us assume that the designer intervenes in the process by assigning a degree of relevance for each solution according to his/her needs. Let us denote as $d_i^j$ the degree of relevance that has been declared by the $i$-th user for $j$-th solution of a particular description. We assume that each solution is represented through a feature vector, say $\mathbf{f}$, which includes the properties of the solution in the design environment. Let us consider that a user, for the example the $i$-th evaluates K solutions, assigning for each of them a degree of relevance, $d_i^{j}$, $j = 1,2,..., K$. Then, we can construct a vector $\mathbf{d}_i$, which includes the relevance degrees of all $M$ solutions as have been assigned by the $i$-th user. That is,

$$\mathbf{d}_i = [d_i^1, d_i^2 ...., d_i^K]^T \tag{9.1}$$

Another user evaluates the same solutions in a different manner, resulting in a different relevance vector the $\mathbf{d}_j = [d_j^1, d_j^2 ...., d_j^K]^T$. In case the two designers have evaluated a different number of solutions, we assume that the relevance degree of the non-evaluated objects is zero.

Then, we can compute the consistency between the $i$-th and $j$-th designer, say $c_{i,j}$, as the absolute correlation coefficient between the relevance vectors of the two designers.

$$c_{i,j} = \rho_{i,j} \tag{9.2}$$

where with $\rho_{i,j}$, we denote the correlation coefficient between the two vectors $\mathbf{d}_i$ and $\mathbf{d}_j$ respectively.

$$\rho_{i,j} = \frac{\mathbf{d}_i^T . \mathbf{d}_j}{\sqrt{\mathbf{d}_j^T . \mathbf{d}_j} \sqrt{\mathbf{d}_i^T . \mathbf{d}_i}} \tag{9.3}$$

The concept behind this heuristic adopted in equation (8.2), as indicator of the consistency between two designers is the following. Small values of $c_{i,j}$ correspond to small values of the coefficient $\rho_{i,j}$, meaning that the degree of relevance assigned by the two designers $i$ and $j$ is almost the same. On the other hand, large values of $c_{i,j}$ means that $\rho_{i,j}$ is large indicated that the two designers are inconsistent for the same design.

### 9.4.1 Problem Formulation

Using values $c_{i,j}$, we can cluster the designers in a way that all designers in a cluster present the maximum consistency among them, that is they present almost the same evaluation for the retrieved objects.

Let us denote as $C_r$ a set which contains designers of the same profile, that is designers of small values of $c_{i,j}$. In this case, we have that

$$Q_r = \frac{\displaystyle\sum_{i \in C_r, j \in C_r} c_{ij}}{\displaystyle\sum_{i \in C_r, j \in V} c_{ij}} \qquad (9.4)$$

where $V$ which includes all the available designers.

The denominator of equation (9.4) expresses the consistency values for the designers assigned in a cluster and all the available one. The denominator is used for normalization purposes to avoid a trivial solution of the assignment of one designer to a cluster.

Equation (9.4) indicates a measure for the Intra-cluster consistency. In the same way we can define the Inter cluster consistency as follows

$$P_r = \frac{\displaystyle\sum_{i \in C_r, j \notin C_r} c_{ij}}{\displaystyle\sum_{i \in C_r, j \in V} c_{ij}} \qquad (9.5)$$

Assuming that the designers are grouped into $M$ clusters, we can take the effect of the $M$ clusters by summing equations (9.4) and (9.5). Therefore, we have that

$$Q = \sum_{r=1}^{M} Q_r \qquad (9.6)$$

and

$$P = \sum_{r=1}^{M} P_r \qquad (9.7)$$

However, it is clear that the

$$P + Q = M \qquad (9.8)$$

Equation (9.8) means that *maximization* of $Q$ [see equation (9.6)] simultaneously yields a minimization of $P$ [see equation (9.7)] and vice versa. Hence, in our problem the two aforementioned optimization requirements are in fact identical and they can be satisfied simultaneously. Therefore, it is enough to optimize (maximize or minimize) only one of the two criteria. In our case, and without loss of generality, we select to minimize $P$, estimating an optimal designers' assignment on the $M$ clusters. Therefore, we have that

$$\hat{C}_r : \min P = \min \sum_{r=1}^{M} \frac{\displaystyle\sum_{i \in C_r, j \notin C_r} c_{ij}}{\displaystyle\sum_{i \in C_r, j \in V} c_{ij}} \text{, for all } r=1,\dots,M \qquad (9.9)$$

In equation (9.9), we define as $\hat{C}_r$ the optimal cluster partition.

## 9.5  Profile Estimation Using Spectral Clustering

Optimizing equation (9.9) is actually a NP-complete problem. Even for the special case of two clusters, i.e., $M=2$, the optimization of (9.9) is practically impossible to be implemented for large number of designers. However, we can overcome this difficulty by transforming the problem of (9.9) into a matrix based representation. Then, an approximate solution in the discrete space can be found using concepts derived from eigenvalue analysis.

### 9.5.1  Matrix Representation

Let us denote as $\Sigma = [c_{ij}]$ a matrix which contains the values of the consistency measures $c_{ij}$ for all $N \times N$ combinations ($N$ is the number of designers) between two designers $i$ and $j$. Let us now denote as $\mathbf{e}_r = [\cdots e_r^u \cdots]^T$ an $N \times 1$ *indicator vector* whose elements $e_r^u$ are given by,

$$e_r^u = \begin{cases} 1 \text{ if the designer } u \text{ is assigned on cluster } r \\ \quad 0 \qquad\qquad\qquad\qquad\qquad \text{Otherwise} \end{cases} \qquad (9.10)$$

In other words, the indicator vector $\mathbf{e}_r$ points out which of the $N$ available designers are assigned on the $r$-th cluster. That is, designers assigned on cluster $r$ are marked with one, while the remaining indices take zero values. Since we have assumed that the M clusters are available, $M$ different indicator vectors $\mathbf{e}_r$, $r = 1, 2, ..., M$ exists.

Therefore, the optimization problem of (9.9) is equivalent to the estimation of the optimal indicators vectors $\hat{\mathbf{e}}_r$, $\forall r$. Consequently, equation (9.9) can be written as

$$\hat{\mathbf{e}}_r, \forall r : \min P = \min \sum_{r=1}^{M} \frac{\displaystyle\sum_{i \in C_r, j \notin C_r} \sigma_{ij}}{\displaystyle\sum_{i \in C_r, j \in V} \sigma_{ij}} \qquad (9.11)$$

The main difficulty in (9.11) is that its right part is *not expressed* as a function of the indicator vectors $\mathbf{e}_r$. Thus, direct minimization is not straightforward.

Consequently, we need to re-write the right part of equation (9.11) in a form of vectors $\mathbf{e}_r$. For this reason, let us denote as $\mathbf{L} = diag(\cdots l_i \cdots)$ a diagonal matrix, whose elements $l_i$, $i=1,2,..N$ express the cumulative consistency degree of a designer with all the remaining ones.

That is,

$$l_i = \sum_j \sigma_{ij} \qquad (9.12)$$

Using matrices $\mathbf{L}$ and $\Sigma$, we can express the numerator of (9.11) as a function of vectors $\mathbf{e}_r$. In particular,

$$\mathbf{e}_r^T (\mathbf{L} - \Sigma)\mathbf{e}_r = \sum_{i \in C_r, j \notin C_r} \sigma_{ij} \tag{9.13}$$

In a similar way, the denominator of (9.11) is related with the indicator vector $\mathbf{e}_r$ as follows,

$$\mathbf{e}_r^T \mathbf{L} \mathbf{e}_r = \sum_{i \in C_r, j \in V} \sigma_{ij} \tag{9.14}$$

Using (9.13) and (9.14), we can re-write (9.11) as

$$\hat{\mathbf{e}}_r, \forall r : \min P = \min \sum_{r=1}^{M} \frac{\mathbf{e}_r^T (\mathbf{L} - \Sigma)\mathbf{e}_r}{\mathbf{e}_r^T \mathbf{L} \mathbf{e}_r} \tag{9.15}$$

Equation (9.15) is the formula that yields the optimal vectors $\hat{\mathbf{e}}_r$, that is those vectors that minimize the consistency among the designers assigned to different clusters, while consequently provide a maximization of the designers consistency within a cluster. Due to (9.8), minimization of (9.15) is equivalent to the maximization of (9.6). Thus, the optimal vectors $\hat{\mathbf{e}}_r$ also return a clustering policy that maximizes the overall consistency for each cluster.

### 9.5.2  Optimization in the Continuous Domain

The main problem in the solution derived from the optimization algorithm described in section 9.5.3 is that the optimal indicators vectors should be in discrete values, since each designer is assigned only to one cluster. Thus, the indicator vectors $\mathbf{e}_r$ take values of one for tasks executed on the $r^{th}$ processor and values of zero otherwise. In other words, if we form the indicator matrix $\mathbf{E} = [e_1 \cdots e_M]$, the columns of which refer to the $M$ different clusters, while the rows to the $N$ available designers. Then, the rows of $\mathbf{E}$ have only one value equal to one while all the rest ones are zero.

Optimization of (9.15) under the discrete representation of the indicator matrix $\mathbf{E}$ (or equivalently the indicator vectors $\mathbf{e}_r$) is still a NP hard problem. However, if we relax the indicator matrix $\mathbf{E}$ to take values in continuous domain, then we can solve the problem in polynomial time. We denote as $\mathbf{E}_R$ the *relaxed version of the indicator matrix* $\mathbf{E}$, i.e. a matrix where its rows take real values instead of indicator matrix $\mathbf{E}$, where its rows have only one value equal to 1 while the rest are 0 (binary values). Then, we discretize the continuous values of the relaxed matrix $\mathbf{E}_R$ to get an approximate solution of the scheduling problem.

It can be proven that the right part of equation (9.15) can be written as [17]

$$P = M - trace(\mathbf{Y}^T \mathbf{L}^{-1/2} \mathbf{\Sigma} \, \mathbf{L}^{-1/2} \mathbf{Y}) \tag{9.16a}$$

Subject to $\mathbf{Y}^T \mathbf{Y} = \mathbf{I}$ (9.16b)

where $\mathbf{Y}$ is a matrix which is related with the matrix $\mathbf{E}_R$ through the following equation

$$\mathbf{L}^{-1/2} \mathbf{Y} = \mathbf{E}_R \mathbf{\Lambda} \tag{9.17}$$

In (9.17), $\mathbf{\Lambda}$ is any abritrary $M \times M$ matrix. In this paper, we select $\mathbf{\Lambda}$ to be equal to the identity matrix, $\mathbf{\Lambda} = \mathbf{I}$. Then, the relaxed indicator matrix $\mathbf{E}_R$, which is actually the matrix we are looking for, is given as

$$\mathbf{E}_R = \mathbf{L}^{-1/2} \mathbf{Y} \tag{9.18}$$

Minimization of (9.16) is obtained through the Ky-Fan theorem 0. The Ky Fan theorem states that the maximum value of the $trace(\mathbf{Y}^T \mathbf{L}^{-1/2} \mathbf{\Sigma} \, \mathbf{L}^{-1/2} \mathbf{Y})$ with respect to the matrix $\mathbf{Y}$, subject to the constraint that $\mathbf{Y}^T \mathbf{Y} = \mathbf{I}$ is equal to the sum of the $M$ ($M<N$) *largest eigenvalues of matrix* $\mathbf{L}^{-1/2} \mathbf{\Sigma} \, \mathbf{L}^{-1/2}$. Thus,

$$\max\{trace(\mathbf{Y}^T \mathbf{L}^{-1/2} \mathbf{\Sigma} \, \mathbf{L}^{-1/2} \mathbf{Y})\} = \sum_{i=1}^{M} \lambda_i \tag{9.19}$$

where $\lambda_i$ refers to the $i^{th}$ largest eigenvalue of matrix $\mathbf{L}^{-1/2} \mathbf{\Sigma} \, \mathbf{L}^{-1/2}$.

However, maximization of (9.19) leads to minimization of $P$ in (9.16a). Thus, it is clear that the *minimum value* of $P$ is given as

$$\min P = M - \sum_{i=1}^{M} \lambda_i \tag{9.20}$$

The Ky-fan Theorem also states that this minimum value of $P$ is obtained for the matrix

$$\mathbf{Y} = \mathbf{U} \cdot \mathbf{R} \tag{9.21}$$

where $\mathbf{U}$ is a $N \times M$ matrix the columns of which are the *eigenvectors* of the $M$ largest eigenvalues of matrix $\mathbf{L}^{-1/2} \mathbf{\Sigma} \, \mathbf{L}^{-1/2}$ and $\mathbf{R}$ an *arbitrarily rotation matrix* (i.e., orthogonal with determinant of one). Again, a simple approach is to select matrix $\mathbf{R}$ as the identity matrix, i.e., $\mathbf{R} = \mathbf{I}$, that is

$$\mathbf{Y} = \mathbf{U} \tag{9.22}$$

Therefore, the optimal values $\hat{\mathbf{E}}_R$ for the relaxed matrix $\mathbf{E}_R$ in the continuous domain will be given as

$$\hat{\mathbf{E}}_R = \mathbf{L}^{-1/2}\mathbf{U} \tag{9.23}$$

Equation (9.23) means that the optimal relaxed matrix is related by i) the cumulative non-overlapping degree of a task with all the remaining ones and ii) the eignevcetors of the $M$ largest eigenvalues of the matrix $\mathbf{L}^{-1/2}\boldsymbol{\Sigma}\,\mathbf{L}^{-1/2}$.

The computational complexity involved in (9.23) is mainly due to the eigenvalue/eigenvector estimation, which is of $O(M^3)$, that is of polynomial order.

### 9.5.3  Discrete Approximation

The optimal matrix $\hat{\mathbf{E}}_R$ given by equation (9.23) has not the form of the indicator matrix $\mathbf{E}$ since the values of $\hat{\mathbf{E}}_R$ are continuous, while $\mathbf{E}$'s rows present only one element of value 1 while the rest ones are zeros. Consequently, the problem is how to round the continuous values of $\hat{\mathbf{E}}_R$ in a discrete form that approximates matrix $\mathbf{E}$.

One simple solution, regarding the rounding process, is to set the maximum value of each row of matrix $\hat{\mathbf{E}}_R$ to be equal to 1 and let the remaining values to be 0. However, such an approach yields unsatisfactory performance in case that there is no a dominant maximum value for each row of $\hat{\mathbf{E}}_R$. Furthermore, it handles the rounding process as $N$ (equal to the number of rows of $\hat{\mathbf{E}}_R$, that is the number of designers) independent problems. An alternative approach, which is adopted in this paper, is to treat the $N$ rows of matrix $\hat{\mathbf{E}}_R$ as $M$-dimensional feature vectors. The algorithm clusters each row of matrix $\hat{\mathbf{E}}_R$ to $M$ clusters as the number of available designers. The rows of $\hat{\mathbf{E}}_R$ indicate the degree of "fitness" (or equivalently the association degree) of a designer to all the $M$ clusters. Therefore, the goal of the clustering algorithm is to find the most appropriate partitions to which a designer with a specific feature vector best fit.

In particular, initially we normalize the rows of matrix $\hat{\mathbf{E}}_R$. Then, we apply the k-means clustering algorithm to these $N$ vectors to form the indicator matrix $\mathbf{E}$. The k-means algorithm composes of three phases, the initialization, the clustering construction, and the updating phase.

*Initialization:* In this phase, the algorithm arbitrarily selects a set of row vectors of $\hat{\mathbf{E}}_R$ as centers of the clusters that are to be constructed. The number of classes equal to $M$.

*Clustering Construction:* In this phase, all the remaining vectors of $\hat{\mathbf{E}}_R$ are assigned to the $M$ clusters using a metric distance. In particular, a vector is assigned to a cluster by comparing this vector with all cluster centers and selecting as appropriate cluster the one with the closest center to the chosen vector.

*Updating:* After the classification, new centers are created as the average value of all vectors belonging to a cluster. In case that these centers are different from the previous ones, a new process takes place and the algorithm moves on the clustering construction phase for further processing. On the contrary, if the new centers are exactly the same with the previous ones, meaning that the same task assignment have been concluded, no further processing is required and the algorithm is terminated.

The performance of the k-means algorithm highly depends on the initial selection of the cluster centers, although it can be proven that the k-means always converges to a solution. Thus, the effectiveness of the schema is actually influenced by the initial selection of the cluster centers in the initialization phase. In this paper, to overcome such a drawback and simultaneously to search for new possible solutions that will yield, in relatively small time, a satisfactory approximation of the optimal solution in the discrete domain, we repeat the experiment by selecting each time different vectors in the initialization phase, which, in the sequel, will provide different solutions. Among all selections, the minimum is returned as the closest approximation.

### 9.5.4  Cluster Integration

As representative user, we select the user of whom the respective feature vector is located nearest to the center of the cluster. Let us denote as $r_k$ the representative designer of the cluster $k$.

After the $k$ representatives have been obtained, each is assigned a weight $w_k$ which is calculated as the sum of two partial weights: a custom index $f_k$ extracted from system use and seniority of the $k$-th user, and the density $d_k$ of the $k$-th cluster which the corresponding user represents. Formally,

$$w_k = a \cdot f_k + b \cdot d_k \tag{9.24}$$

where $a$, $b$ represent constant factors, common for all designers, allowing to fine tune the contribution of each partial weight to the result. The final weights are provided to the system through the *members' importance* vector. In the following we assume that $w_k \geq 1$ without loss of generality.

Finally, for each solution $s_i$ the overall evaluation taking into account the entire group of representatives is calculated as a simple sum, i.e.

$$e_i = \sum_{m=1}^{k} e_{i,m} w_m \tag{9.25}$$

where $e_{i,m}$ is the evaluation of solution $s_i$ by representative $m$. Solutions included in the final consensus set fulfil the simple condition of

$$e_i \geq T \tag{9.26}$$

where T represents an adjustable threshold for each submitted description, which may be used to control the consensus set population.

## 9.6 Experimental Results

In this section we present experiments with concern of the proposed algorithm. In addition, we compare our scheme with other methods for filtering the generated solutions. Section 9.6.1 presents the experimental environment setup, while section 9.6.2 the simulation outcomes. In addition, in section 9.6.1, we describe the objective evaluation criteria used for measuring the performance the proposed algorithm and comparing it with the other methods.

### 9.6.1 Experiments Setup

To setup the experiments, we use a large set of different scene descriptions. As we have previously stated, for each scene description a large set of generated solutions are derived. This is due to the fact that usually a scene is described under an intuitive manner, meaning that the description is often imprecise. The generated solutions are then evaluated by the users and a degree of relevance/ irrelevance is assigned to each solution. By exploiting these degrees of relevance/irrelevance, our proposed algorithm groups the users into clusters of similar preferences.

***Approaches that they have been implemented:*** Apart form the proposed algorithm three other implementations have been used for testing the system and comparing it. The first approach concerns a solution-based filtering approach which it does not take into consideration user's collaboration. We refer this method as *No Collaborative Preference Consensus,* since user's profile is estimated using information coming from a single isolated user.

The second approach, called *User Preference Consensus*, actually implements the approach of [16]. In particular, in this framework we estimate the profile of a user by taking into account the evaluation of all the collaborative users involved in a design.

Finally, the third implementation is almost the same with the proposed algorithm with the difference that another clustering method is adopted. More specifically, in this implementation, we perform user preferences clustering through the use of the k-means mechanism. The *k*-means is a very well known and stable clustering method, with, however, the drawback that classification is accomplished in a centric-based scheme. This means that two similar points may categorize into different classes if the distance of the first point with respect to the center of one class is smaller than the distance with respect to the other class, while the opposite happens with the second point. We call this approach *k-means clustering* in the following of this paper.

To overcome this difficulty, an intelligent clustering mechanism is implemented in this paper by taking into consideration the relation degree between two users.

***Evaluation Criteria:*** The efficiency of the algorithm is measured using objective criteria, such as the precision-recall curve. Precision and Recall are two widely used measures for evaluating the quality of results in domains such as Information Retrieval. Precision can be seen as a measure of *exactness* or *fidelity*, whereas Recall is a measure of *completeness*. In particular, Precision is defined as the number of relevant returned solutions divided by the total number of solutions retrieved. It is clear that high values of precision imply better filtering algorithms compared. Instead, Recall is defined as the number of relevant solutions retrieved divided by the total number of existing relevant solutions (which should have been retrieved).

Mathematically, precision is defined using the following equation

$$P_r = \frac{\# \text{number of relevant retrieved solutions}}{\text{Total number of retrieved solutions}} \tag{9.27a}$$

Instead, recall is measures using the following quantity

$$Q_r = \frac{\# \text{number of relevant retrieved solutions}}{\text{Total number of relevant solutions}} \tag{9.27b}$$

A perfect Precision score of 1.0 means that every result retrieved by a search was relevant (but says nothing about whether all relevant documents were retrieved) whereas a perfect Recall score of 1.0 means that all relevant documents were retrieved by the search (but says nothing about how many irrelevant documents were also retrieved). For this reason, the precision-recall curve is usually used for measuring the efficiency of retrieval or filtering algorithm.

Another criterion for measuring the performance of an information retrieval system is the Average Normalized Modified Retrieval Rank, ANMRR. The ANMRR hae been introduced by the MPEG-7 standard for measuring the performance of a content-based retrieval system [19]. The advantage of the ANMRR compared with the precision-recall curve is that it measures the ranking of the precise data, a rate that it lost in the precision criterion. ANMRR takes value in the between zero and one. Values close to zero indicates the perfect matching, while values around one the worst matching.

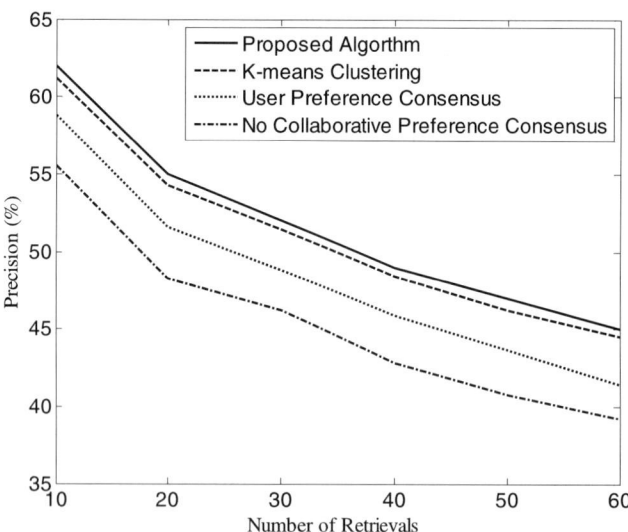

**Fig. 9.3.** Precision versus the number of solutions returned to the end user for the proposed algorithm and three other implementations

### 9.6.2 Simulations

In this section, we present simulation results as far as the proposed clustering algorithm is concerned and the aforementioned described implementation schemes. Initially, we evaluate the precision measure at different number of retrievals. Fig. 9.3 depicts the variation of the precision measure versus the number of the returned solutions. In this figure, we have compared the proposed algorithm with the three implementations described in section 9.6.1 for solution filtering.

As is expected, as the number of the returned solutions increases, the precision accuracy decreases. However, the proposed method outperforms the other three approaches providing better precision results at a given number of solution-retrievals. The worst method is the one where no user collaboration is taken into consideration.

Fig. 9.4 presents the precision-recall curve for the proposed method and the other three implementations. Again, we observe that our scheme outperforms the examined ones. The worst performance is also noticed for the method of [10], where no user collaboration is taken into account for the solution filtering.

Table 9.1 presents the ANMRR values for the proposed algorithm and the three different implementation approach presented in section 9.6.1, i.e., the *k-means*

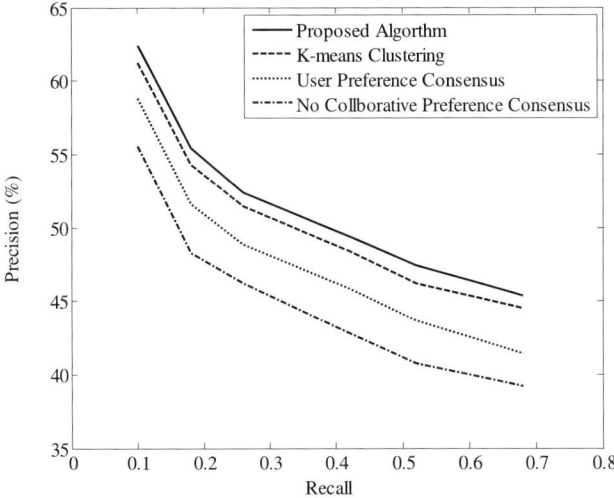

**Fig. 9.4.** The Precision-Recall curve for the proposed method and the other three implementations

**Table 9.1.** The SER ratio of the proposed scheme compared with other approaches

| Average Normalized Modified Retrieval Rank- ANMRR | | | |
|---|---|---|---|
| The Proposed Algorithm | K-means Clustering | User Preference Consensus | Non-Collaborative Preference Consensus |
| 0.18 | 0.23 | 0.34 | 0.45 |

*clustering*, the *user preference consensus* and the *non-collaborative preference consensus* method. As we observed, the best performance (smallest ANMRR value) has the proposed algorithm, while the worst has the method of [10], where the user profile is estimated with non collaborative way. Finally, Table 9.2 presents the precision values for different number of clusters. As we observed, small number of user profile partitions deteriorates the results. This is due to the fact that in this case, we group together users of different.

## 9.7 Conclusions and Future Work

In this work we present an approach towards solution evaluation in the context of a collaborative declarative design environment taking advantage of user information on previous system use and additional custom user information. The mechanism employed for this purpose is based on user clustering with respect to preferences and subsequent extraction of a representative for each cluster. The evaluation of these representatives for newly generated solutions are employed for the construction of the final solution set that will be yielded as the consensus of the team that has collaborated for the declarative description definition.

Our approach improves the efficiency of a collaborative declarative design environment by restricting the set of users actively participating in the collective solution evaluation while ensuring adequate representation of all trends in user preferences due to the construction of the designer clusters. Moreover, it allows control over the importance of each designer's preferences as well as over the strictness imposed upon the solutions comprising the final consensus set.

Future works are the acceleration of the computations times for eigenvalue estimation using either Random Walks scenarios on Graphs or approximate solutions. Additionally, other clustering schemes can be included

## References

[1] Rodriguez, K., Al-Ashaab, A.: A Review of Internet based Collaborative Product Development Systems. In: Proceedings of the International Conference on Concurrent Engineering: Research and Applications, Cranfield, UK (2002)

[2] Choo, W.C., Detlor, B., Turnbull, D.: Web Work: Information Seeking and Knowledge Work on the World Wide Web. Springer, Heidelberg (2000)

[3] Shen, W.: Web-based Infrastructure for Collaborative Product Design: An Overview. In: 6th International Conference on Computer Supported Cooperative Work in Design, pp. 239–244, Hong Kong (2000)

[4] van den Berg, E.: Collaborative Modelling Systems, Technical Report Delft University (October 1999)

[5] Kvan, T.: Collaborative design: What is it? Automation in Construction 9(4), 409–415 (2000)

[6] Wang, L., Shen, W., Xie, H., Neelamkavil, J., Pardasani, A.: Collaborative Conceptual Design –State of the Art and Future Trends. Computer – Aided Design 34, 981–996 (2002)

[7] Plemenos, D.: Contribution à l'étude et au Développement des Techniques de Modélisation, Génération et Visualisation de Scènes – Le projet MultiFormes. Thèse de Doctorat d'Etat, Nantes, (November 1991)

[8] Dragonas, J., Makris, D., Lazaridis, A., Miaoulis, G., Plemenos, D.: Implementation of Collaborative Environment in MultiCAD Declarative Modelling System. In: International Conference on Computer Graphics and Artificial Intelligence (3IA 2005), Limoges (France), May 11-12 (2005)

[9] Plemenos, D., Tamine, K.: Increasing the Efficiency of Declarative Modeling. Constraint Evaluation for the Hierarchical Decomposition Approach. In: International Conference Computer Graphics, Visualization and Computer Vision, Plzen, Czech Republic (1997)

[10] Bardis, G., Miaoulis, G., Plemenos, D.: Design and Configuration of a Machine Learning Component for User Profiling in a Declarative Modelling Environment. In: International Conference on Computer Graphics and Artificial Intelligence (3IA 2004), Limoges (France) (May 2004)

[11] Dragonas, J.: Modélisation Déclarative Collaborative. Thèse de Doctorat, Université de Limoges, France (2006)

[12] Bardis, G.: Apprentissage et aide a la décision pour la modélisation déclarative de scenes. Thèse de Doctorat, Université de Limoges, France (2006)

[13] Plemenos, D.: Declarative modelling by hierarchical decomposition. The actual state of the MultiFormes project. In: International Conference GraphiCon 1995, St Petersburg, Russia (1995)

[14] Herrera, F., Herrera-Viedma, E., Verdegray, J.L.: A Model of Consensus in Group Decision-making under Linguistic Assessments. Fuzzy Sets and Systems 78, 73–87 (1996)

[15] Choudhurya, A.K., Shankarb, R., Tiwaria, M.K.: Consensus-based Intelligent Group Decision-making model for the Selection of Advanced Technology. Decision Support Systems 42, 1776–1799 (2006)

[16] Bardis, G., Doulamis, N., Dragonas, J., Miaoulis, G., Plemenos, D.: A Parametric Mechanism for Preference Consensus in a Collaborative Declarative Design Environment. In: 10th Infographie Interactive et Intelligence Artificielle (3IA 2007), pp. 19–27, Athens, Greece (2007) ISBN 2-914256-09-4

[17] Ng, A.Y., Jordan, M.I., Weiss, Y.: On spectral clustering: analysis and an algorithm. Neural Information Processing Systems 14 (2002)

[18] Nakic, I., Veselic, K.: Wielandt and Ky-Fan Theorem for Matrix Pairs. Linear Algebra and its Applications 369(17), 73–77 (2003)

[19] MPEG-7 Visual part of eXperimentation Model Version 2.0, MPEG-7 Output Document ISO/MPEG (December 1999)

# 10

# OmniEye: A Spherical Omnidirectional Vision System to Sonify Robotic Trajectories in the AURAL Environment

Artemis Moroni[1], Sidney Cunha[1], Josué Ramos[1], and Jônatas Manzolli[2]

[1] Robotics and Computer Vision Division - Renato Archer Research Center
DRVC/CenPRA
Rod. D. Pedro I, km 143,6 – Campinas - São Paulo - 13069-901 - Brazil
[2] Interdisciplinary Nucleus for Sound Studies – University of Campinas
NICS/UNICAMP

**Abstract.** This paper describes the OmniEye, an omnidirectional vision system developed to track mobile robots in AURAL, a computational structured environment. AURAL aims to control the interaction between visual, sound and robotic information in a research for automatic and semi-automatic processes of artistic production. Different convex mirrors can be used to achieve an omnidirectional system. The use of a spherical mirror in this case introduces distortions in the image. A toolbox for the calibration of central omnidirectional cameras was used to obtain a first estimation for the imaging function. On a second step, a genetic algorithm was applied to adjust the coefficients of the imaging function. Experimental results and the application of the OmniEye for translating robotic paths into sound events in the AURAL environment are described.

## 10.1  Introduction

This development is part of the AURAL project, where a user draws a path in an interactive interface and transmits it to a mobile robot. The AURAL explores the *arTbitrariness*, a research for automatic and semi-automatic processes of artistic production [1]. In this case the arTbitrariness occurs in the sound domain. By observing the behavior of a mobile robot in a physical space, the AURAL organizes a sequence of sound events. The robot tries to travel along the path in a follow up area, but it can be disturbed by other robots or obstacles while traversing it. The interaction of physical parameters and the presence of the mechanical bodies of the robots are potentially able to generate a complex sequence of interactive events. These events will be used to modify the performance controls of JaVOX, an evolutionary environment applied to sound production [2]. Like others, AURAL belongs to that kind of systems that combine the behavior of mobile robots with sound events [3, 4, 5]. The OmniEye [6] occupies an important role in AURAL, for it is the "observer" being used to feedback the robot localization. In addition, it is possible to compound a data fusion for other specific tasks such a map generation.

A possible approach to improve the accuracy of camera motion estimation from image sequences is the use of an omnidirectional camera which combines a

D. Plemenos, G. Miaoulis (Eds.): Arti. Intel. Techn. for Comp. Graph., SCI 159, pp. 159–174.
springerlink.com                                    © Springer-Verlag Berlin Heidelberg 2009

conventional camera with a convex mirror that can magnify the field of view [7]. Omnidirectional visual systems provide images with field of view having 360°x180° (horizontal x vertical). Yagi [8] compiled a literature review which shows the application of various types of omnidirectional visual systems. Such systems can be assembled, according to various models, using multiple cameras, which point to different directions, or even using a single free camera which rotates around a fixed axis [9].

Different convex mirrors can be used to achieve an omnidirectional system. Parabolic, hyperbolic, spherical mirrors or even pre-designed surfaces with specific desired properties can be applied. From the possible ways of building an omnidirectional system, it was decided to assemble a spherical mirror. In spite of the fact that the spherical mirror does not present any special property [10], it is relatively easy to be built and can also be used in robotic navigation and tele-operation, having low cost compared to hyperbolic mirrors.

Next, in sections 10.2 and 10.3, a description of the geometrical modeling for the development and calibration of the omnidirectional camera will be made. In section 10.4 the genetic approach to optimize the calibration function will be commented. In section 10.5 some results will be presented and section 10.6 approaches the translation of trajectories into sound events. Finally, in section 10.7, the conclusions.

## 10.2   Geometrical Modeling for an Omnidirectional System

In recent years, new calibration techniques have been developed which can be applied to any kind of central omnidirectional cameras. For instance, Micusik and Pajdla [11] extended the geometric distortion model and the self-calibration procedure, including mirrors, fish-eye lenses and non-central cameras. In [12], the authors describe a method for central catadioptric cameras using geometric invariants. They show that any central catadioptric system can be fully calibrated from an image having three or more lines. In [13], a unified imaging model for fisheye and catadioptric cameras is presented. Finally, in [14], the authors presented a general imaging model which encompasses most projection models used in computer vision and photogrammetry. They also introduced a theory and algorithms for a generic calibration concept.

All these papers can be classified into two main categories. The first one includes methods which explore prior knowledge about the scene [15]. The second group covers techniques that do not use this knowledge. This includes calibration methods from pure rotation or planar motion of the camera [15, 16] and self-calibration procedures performed from point correspondences and epipolar constraints by minimizing an objective function [11]. All these allow the obtention of accurate calibration results, but primarily focus on particular sensor types (e.g. hyperbolic and parabolic mirrors or fish-eye lenses). Moreover, some of them require special settings of the scene and ad-hoc equipment.

This paper belongs to the second group. Here, the development of a geometrical formulation for images aims to determine a relation between the coordinates of the physical world and the coordinates of the pixels of a corresponding omnidirectional image. For this, given a captured image and the corresponding scene of the world, the modeling of the geometrical projection of this image is necessary to relate some measurements of interest.

The use of a spherical mirror, with no single effective viewpoint in the omnidirectional system, introduces distortions in the captured image. The processing of this kind of system can be carried out in two different ways: through an initial rectification of the image and further application of concerning techniques, or the handling of the omnidirectional image. The former approach is useful when the final result of the process is oriented to the human interpretation of the image, while the latter avoids the need of a rectification and can be used for the automatic processing of the image by means of computational systems.

In this paper, the catadioptric omnidirectional system is made up of a camera, a spherical convex mirror and a conical weight, assembled in a pendulum mount, which gives a vertical direction having good accuracy to align the camera and stabilize the set up. To cause a minimal obstruction in the image acquired or, in other words, to obtain regions with minimal occlusions, nylon threads were used to fix the system.

The optical axis of the camera was aligned with the optical center of the mirror, which was hung from the ceiling of the room, minimizing the occlusions. The whole environment was captured in a single image. The set up is shown in Figure 10.1.

### 10.2.1  Intrinsic and Extrinsic Parameters

Specific algorithms to process images acquired with the omnidirectional system invariably require geometrical parameters from the optical system which is being applied. The analysis of the radial distortion introduced by the system camera-mirror with relation to a world scene is of fundamental interest to the spherical mirror, as well as the determination of the intrinsic and the extrinsic parameters of the mathematical model applied to the camera.

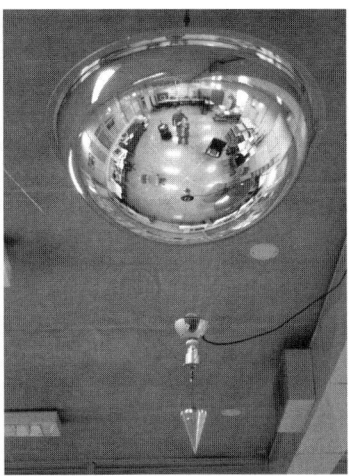

**Fig. 10.1.** The OmniEye with the spherical mirror, the camera fixed with nylon threads and the conical mass that make up the omnidirectional system

The extrinsic parameters are the entries of the translation matrix **T** and the orthogonal matrix **R,** totalizing 6 parameters. The intrinsic parameters are those necessary to determine the optical, geometrical and digital characteristics of the visualization provided by the camera. These parameters can be described by: (1) the geometric projection (characterized by the focal distance **f** of the lens and the pixel size); (2) the transformation of the coordinates of the camera-to-image reference systems and (3) the geometric distortion introduced by the optical system during the process. In spite of the geometric distortion, we have:

$$x_c = -(x_{im} - o_x)s_x; \, y_c = -(y_{im} - o_y)s_y \tag{10.1}$$

where $(x_c, y_c)$ and $(x_{im}, y_{im})$ are the coordinates of the image point of the camera-image reference systems, respectively; $(o_x, o_y)$ are the coordinates of the center of the image and $(s_x, s_y)$ are the actual size of the pixel effective size (in millimeters) in the horizontal and vertical directions, respectively.

### 10.2.2  Omnidirectional Perspective Camera Model

The modeling of the catadioptric omnidirectional perspective camera system permits us to relate the coordinates of the image (in pixels with respect to the axis **u** and **v**) with the three-dimensional vector **p** whose origin is in the unique center of projection of the mirror and whose end is at the point of reference in space. This projection model is based on Scaramuzza et al. [17].

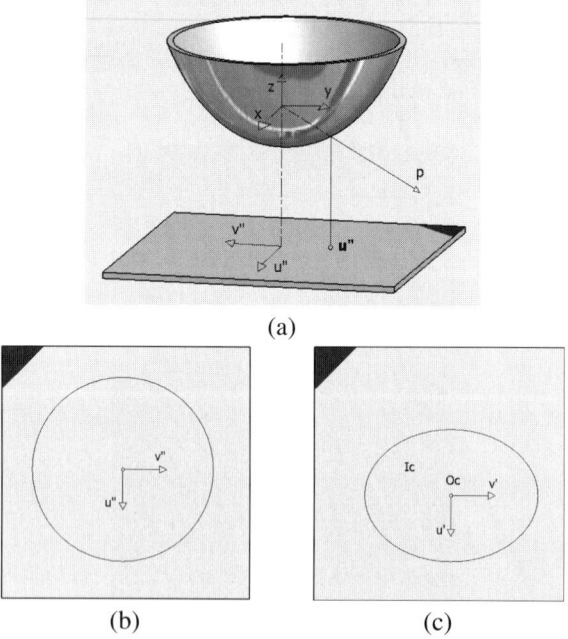

**Fig. 10.2.** (a) The coordinate system in the catadioptric case. (b) Sensor plane, in metric coordinates. (c) Camera image plane, in pixel coordinates. Pictures (b) and (c) are related by an affine transformation.

In spite of the spherical mirror, it is assumed that there is a single center of projection (origin of the vector **p**) which is also the center of the radial symmetry of the mirror with its optical axis. This approximation is assumed because only the central part of the image is actually used in the experiment. The localization of the objects will be made in a restricted area of the environment.

A system with a single center of projection is such that beams of light from the camera, reflected by the surface of the mirror, intersect each other at a single point (virtual point, origin of vector **p** as shown in Figure 10.2). Systems without a single center of projection, in contrast, are those in which the intersection between the described beams do not occur at a single point.

The construction of omnidirectional catadioptric systems employing lenses with hyperbolic, parabolic or elliptic mirrors ensures the property of the single center of projection. For spherical mirrors this property can only be approached locally in the central area of the image. Thus, the coordinate vector **p** = (**x, y, z**) and its projection **u** = (u, v) described in Figure 10.2 can be related as:

$$\begin{bmatrix} x \\ y \end{bmatrix} = \lambda \begin{bmatrix} u'' \\ v'' \end{bmatrix}, \lambda \geq 0 \tag{10.2}$$

$$p = \begin{bmatrix} x \\ y \\ z \end{bmatrix} = \lambda \begin{bmatrix} \alpha \cdot u' \\ \alpha \cdot v' \\ f(\alpha.\rho') \end{bmatrix} \lambda, \alpha \rangle 0 \tag{10.3}$$

Since **p** is a vector, a constant $\alpha$ can be included in f(u, v), since this latter function depends only on the radial distance $\rho^2 = u^2 + v^2$ of point **p** to the optical axis.

$$p = \begin{bmatrix} x \\ y \\ z \end{bmatrix} = \lambda \begin{bmatrix} \alpha.u' \\ \alpha.v' \\ f(\alpha.\rho') \end{bmatrix} \lambda, \alpha > 0 \tag{10.4}$$

Therefore the process of calibration consists of determining the coefficients of the polynomial expression (Taylor series), the intrinsic parameters given below, as well as the extrinsic parameters.

$$f(u'') = a_0 + a_1\rho'' + a_2\rho''^2 + a_3\rho''^3 + a_4\rho''^4 + ... \tag{10.5}$$

By applying a spherical coordinate system, we obtain:

$$u = \rho''\cdot\sin(\theta)\cdot\cos(\varphi)$$
$$v = \rho''\cdot\sin(\theta)\cdot\sin(\varphi) \tag{10.6}$$
$$z = f(\rho'') = \rho''\cdot\cos(\theta)$$

If u and v are known, $\varphi$ and $\theta$ can easily be found. u and v are extracted from the pixels of the image. Since $\varphi$ and $\theta$ are known, the coordinates x and y, associated with the u and v pixel coordinates, can be evaluated for any desired plane z. In this case, the plane is the floor of the room, the (x, y) world coordinates of an image can be calculated, and consequently the path of a mobile robot.

## 10.3   The Omnidirectional System Calibration

The toolbox [18] allows the calibration of any central omnidirectional camera or, in other words, cameras having a single center of projection. The calibration is accomplished in two different stages: initially, a set of images containing a chess pattern is captured from different positions and orientations in space. Then, the corners of the pattern are manually determined using the toolbox. The calibration is then automatically calculated by using the obtained data, with the help of a corner detector to improve the accuracy of the data.

Through the camera calibration the relationship between the pixels of the image and the 3D vector can be determined, as well as the origin in the single projection center and the end in the space points projected on the image, as shown in Figure 10.2.

However, even if the property of a single projection center is not exactly verified, the toolbox still provides good results with the calibration. The spherical mirror furnishes the possibility of a good estimation of a hyperbolic mirror in a restricted area of space, in the central part of the image. During the calibration with the toolbox, the degree of the polynomial used to map the pixels of the image, with the corresponding 3D points of the world, is requested. Experience has shown that polynomials of degree 4 are enough to describe the image-world mapping [17] resulting from the optimization of the SSD function. Once the coefficients in Equation 10.5 are determined, a spherical coordinate system was used to find any vector emanating from the omnidirectional image to the world.

The error incurred was of 8% in the minimum "maximum distance" although the result was very consistent. Aiming to obtain a better approximation, a genetic algorithm was applied to optimize the coefficients.

## 10.4   Image Function Estimation with a Genetic Algorithm

The genetic algorithm applied to estimate the calibration function was a variation of the canonical one, as proposed by Holland [19]. The chromosome of each individual of the population is coded in an array of length 8, where the first five elements contain the coefficients of the polynomial defined by Equation 10.5 being estimated.

The sixth element of the array corresponds to the distance of the focus of the mirror. The seventh and eighth elements contain the coordinates x and y of the center of the image. It is worth to remember that the spherical mirror does not have a well defined focus. The values which are being investigated are those which better estimate the mapping from pixels of the image with the points of the world, whose coordinates are known.

### 10.4.1  The Reproduction Cycle

In the first experiment using the genetic algorithm, eight points $P_i$, i = 1 ... 8, were marked on the floor of the room for fitness evaluation. An image of the scene was captured using the omnidirectional system and the pixel coordinates $(u_i, v_i)$ corresponding to each of the eight points were obtained from this image.

The tournament selection was applied to choose the parents for the next generation. The values obtained for each individual from Equation 10.6 were applied as parameters in Equation 10.7, used to estimate position $P_i'$ for each point $P_i$. The distance $d_i$ between each pair $(P_i, P_i')$ was evaluated and D was assigned with the greatest $d_i$. The fitness F for each individual was evaluated as:

$$F = 1/D \tag{10.7}$$

Therefore, what was investigated was the minimum "maximum image-world distance". An arithmetic crossover was applied to the pairs of parents, followed by the Gaussian mutation [20]. The best individuals of the previous generation were included in the new one.

### 10.4.2  The Results

Since the first experiment using the genetic algorithm, the distance was 3% of the mmd (minimum "maximum distance"). It is worth to note here that there are some sources of errors. For example, the alignment of the camera with the center of the mirror in the vertical plane is very sensitive to errors. Aiming to obtain more points to improve the imaging function, a grid of points was drawn on the floor. Figure 10.3 shows the image that was used to obtain the pixels of the points of the grid and Figure 10.4 depicts the mapping.

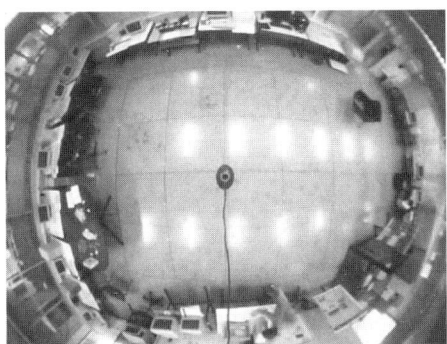

**Fig. 10.3.** The image of the grid, captured with the omnidirectional system

The function fitness was evaluated considering 24 points. A better result was obtained with an error of 1% relative to the maximum distance, 5000 individuals, mutation rate = 15%; crossover rate = 30%. In this application, the result was satisfactory since the robot used in the experiment was a Nomad 200, having a 45 cm diameter and 85cm high. The height of the mirror from the floor was 2.9 m.

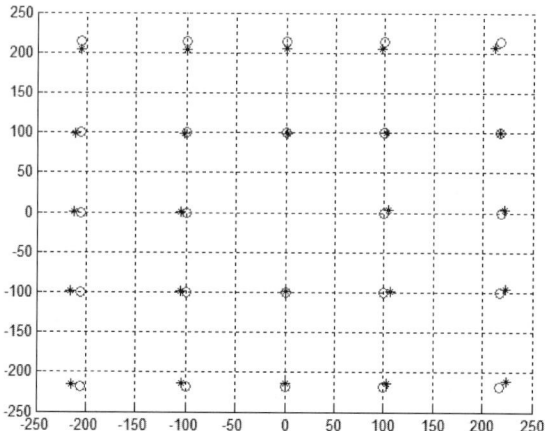

**Fig. 10.4.** The map that was obtained using the genetic algorithm with 24 points to evaluate the fitness. The circles are the points of the world; the crosses are the estimated points.

## 10.5 Tracking the Robot

Functions of the OpenCV library are being used to track the robot. Initially, the vision system captures an image of the environment that will be used as a background image. This image is subtracted from all the other images which were obtained in real time. If no modification of brightness in the environment has occurred, the result of the subtraction is a black image.

On the robot, a high intensity light source was mounted. The lamp is lit after the background image has been captured. Each image captured with the omnidirectional vision system is subtracted from the background image and a thresholding function is applied. The result is a binary image (black and white), according to the threshold level applied. Operations of mathematical morphology (Top Hat, Opening, Closing) are then applied in the binary image. Next, a routine to find contours is used to obtain the location (pixels) in the image of the mark associated with the lamp of the robot. The coordinates of the pixels of the contour are then used to calculate the centroid of the mark. Following, the mapping function is used to evaluate the coordinates of the world associated with the coordinates of the centroid. These coordinates of the world are used as feedback to the robot concerning its position in the environment.

Figures 10.5 (a) and (b) show the results of the morphological operations with the omnidirectional images, pointing out the path traversed by two robots, the Nomad and a Roomba. In (a), a trajectory was sent to be traversed by the Nomad robot. In (b), the way traversed by a Roomba robot is shown. Note that the spiral performed by the Roomba, when turned on, is easily recognized in the upper part of Figure 10.5 (b).

The developed code allowed to accomplish all the mentioned operations in real time recording video images at 30 fps, with all the robot path logs and localizations in the world referential, as seen in Figure 10.5 But the use of a light source for tracking the robot presupposes that there is not a great light variation in the environment. To

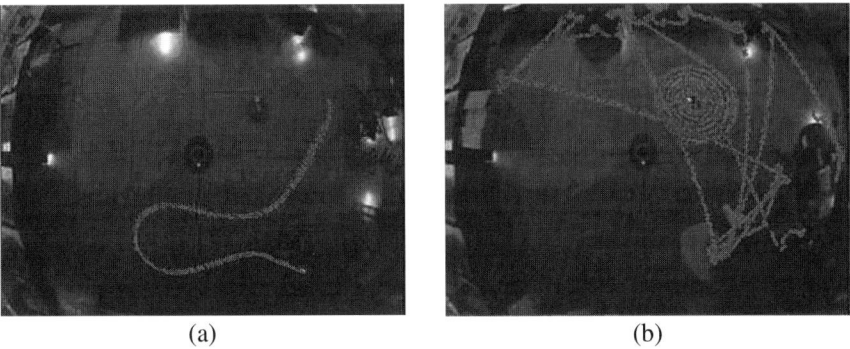

(a)                                              (b)

**Fig. 10.5.** On the left, the path traversed by the Nomad robot, observed by the omnidirectional system. On the right, the path traversed by the Roomba robot.

surpass this limitation, another approach based on colors was applied. On each robot, a strongly-colored panel was fixed and a variation of the Camshift demo from OpenCV samples was applied. In short, once the program is launched, a rectangle on the panel to be tracked is selected with the mouse, in the image captured with the OmniEye. A color histogram is created to represent the panel. Next, the "panel probability" for each pixel in the incoming video frames is calculated. The location of the panel rectangle in each video frame is shifted. The new location is found by starting at the previous location and computing the center of gravity of the panel-probability values within a rectangle. The rectangle is then shifted to its right over the center of gravity.

Camshift stands for "Continuously Adaptive Mean Shift" and is based on Mean Shift algorithm [21]. The algorithm is called *continuously adaptive* and not just *mean*

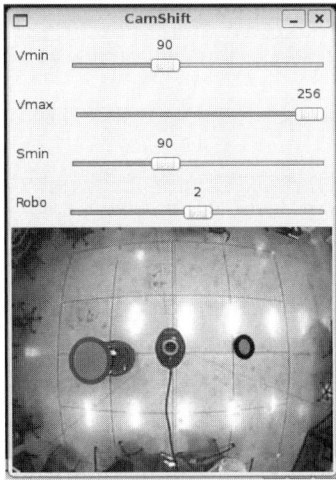

**Fig. 10.6.** On the left, the Nomad robot, inside the red circle; on the right, the Roomba inside the blue circle, both tracked with the CamShift interface

*shift* because it also adjusts the size and angle of the panel rectangle each time it shifts it. It does this by selecting the scale and orientation that are the best fit to the panel-probability pixels inside the new rectangle location.

Figure 10.6 shows the Camshift interface, and the robots Nomad (red circle) and Roomba (blue circle). The coordinates of the center of mass of the circles in the image are evaluated and applied in the equation system 10.6 to evaluate the position of each robot in the region.

## 10.6   Translating Paths into Sound Events

The system that translates paths into sound events is also based on Evolutionary Computation. In this context, the MIDI protocol representation was used to code the genotype, like in the original development of VOX POPULI [22, 23]. This environment, initially developed in Visual Basic, was translated into Java, resulting in JaVOX. The features described in this paper are present both in JaVOX and VOX POPULI.

In AURAL, the robot's paths are associated with sound structures in JaVOX. The paths are transformed into sound events by using populations of clusters and chords of

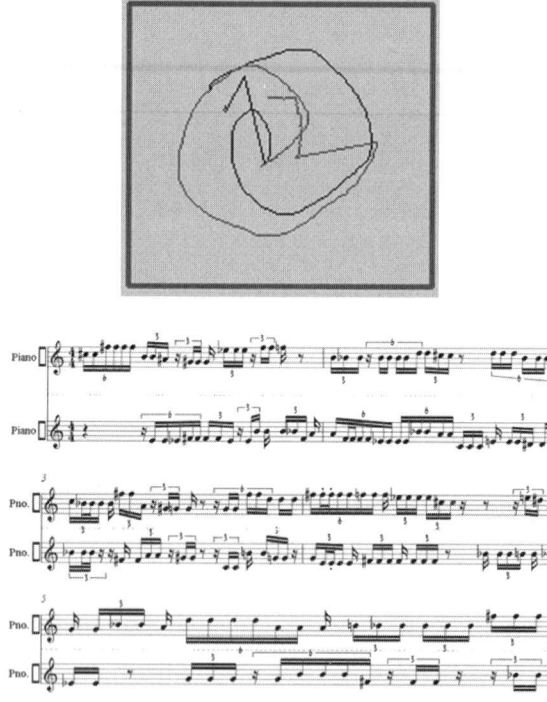

**Fig. 10.7.** Sound sequence resulting from the curves above, drawn in the interactive pad of VOX POPULI environment

four MIDI notes. An analogy can be made of each individual of the population with a choir of four voices that can be classified as *bass, tenor, contralto* and *soprano*. At each step of the process a new sonority is created. The choir with highest fitness is selected in the evolutionary process and sent to the MIDI board to be played.

In both environments, a control area (pad) of the interactive interface enables the user to draw curves in a phase space, associating to each one of them a trajectory that guides the sound production. Figure 10.7 shows the curves drawn by the user in the graphic interface of VOX POPULI and the resulting sound sequence.

Figure 10.8 shows the interface of JaVOX environment, where the lines drawn by the user direct the sound production in real time. Like in VOX POPULI, JaVOX links each line with the interface parameter controls. The red curves are associated with the melodic parameter (mel), in the x-component, and octave parameter (oct), or voices interval, in the y-component. These parameters guide the evolutionary sound production.

Similarly, in AURAL, the paths are drawn and transmitted to a mobile robot. The mobile robot traverses a structured region which is associated, through a bi-dimensional

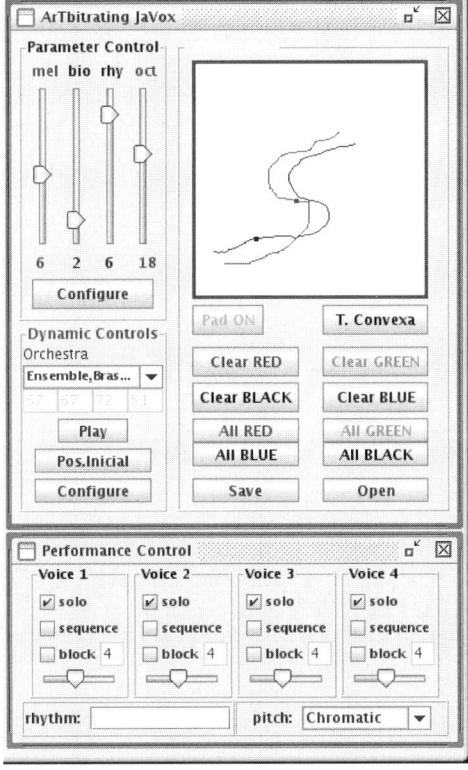

**Fig. 10.8.** The JaVOX interface. Below, the performance controls that will be associated with events decurring from the interaction between the two robots.

projection, with the area in the graphic interface that is approximated with MIDI events. The robot is observed by the OmniEye, that estimates the location of the robot in the area and sends it to JaVOX. The corresponding position is plotted in the interactive pad. The sequence of points describes the approximated path traversed by the robot.

In Figure 10.8, the red line was drawn by the user and sent to the robot as a trajectory to be traversed. The blue line represents the path traversed, observed by the OmniEye. Both are used to control the sound production.

### 10.6.1 Musical Performance Controls

Besides the trajectories, JaVOX has other possibilities to control the sound production in real time. See, in the lower part of Figure 10.8, the Performance Control interface. For each one of the four voices there are three controls named 1) solo; 2) sequence; and 3) block. The performance controls work as delay lines in which individuals from previous generations can be played as solo, melodic patterns or chords. This adds an ecological component in the JaVOX Evolutionary process. Robot's real time behavior is used to reassign individuals from previous generations, recycling note material and playing them in real time. When the solo control is enabled, the sound events are sent direct by the MIDI board from the pad control and the evolutionary process, which supplies an event MIDI sequence at each JaVOX interaction. Therefore, for the first control the sound result depends on the interaction between the curve transmitted to the robot and the curve observed by the OmniEye, that represents the approximated path traversed by the robot.

In the second performance control (sequence), the four MIDI notes (voices) of the cluster (or choir selected in the evolutionary process) are played in sequence. In this

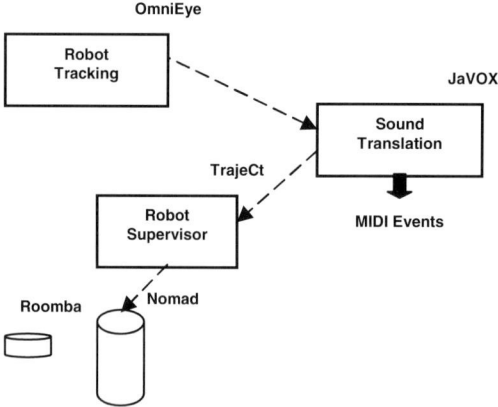

**Fig. 10.9.** The AURAL architecture diagram. The curves drawn on the JaVOX interface are transmitted as trajectories to the mobile robot. The OmniEye observes the paths traversed by the robots and send it to JaVOX, wherein they are plotted as curves. The curves in JaVOX graphic interface control the sound generation.

way the horizontal character of the performance is emphasized, generating a sound texture with a melodic character. In the third control (block), the MIDI events are sent to the MIDI board as fast as possible, almost simultaneously, generating a superposition of note blocks. In this case the emphasis is in the verticality of sound events, generating a complex cluster texture.

These three modes of sound performing generate significant variations in the sound result and can be applied as a compositional strategy. The interaction of these controls with the dynamic behavior of the mobile robot, the OmniEye and eventually, the presence of other robots in the area, can generate a complex sound organization.

The link process between the behavior of the robots in the structured area and the translation into sound was developed aiming to verify the capability of the AURAL to create self organized sound textures departing from simple interactions between the agents of the system, i. e., the mobile robots. A supervisor module TrajeCt (for *traject control*) receives the sequence of trajectory points from JaVOX and sends it to the Nomad. Another mobile robot, a Roomba, moves freely in the area using an autonomous navigation system. When there is a collision, the Roomba moves out. The flow of information departs and returns to JaVOX in the sound translating process, as shown in Figure 10.9.

Communication between each part of the system is made by means of an Interprocess communication (IPC). The path traversed by the mobile robots is captured by

**Fig. 10.10.** In the picture, the OmniEye with three mobile robots: a Roomba, a Pioneer and Nomad

**Table 10.1.** Proximity and Performance Control in AURAL

| Distance between the mobile robots | | | |
|---|---|---|---|
| | Solo | Sequence | Bloco |
| Far | X | | |
| Medium | | X | |
| Close | | | X |

the OmniEye that provides the coordinates (Equation 10.1) and the criteria of behavior for performance control in JaVOX.

The interaction between the free navigation of the Roomba(s) and the path traversed by the Nomad generates a collective behavior between the robots that is used as a performance control in JaVOX. Figure 10.10 shows the omnidirectional system and three mobile robots. There may be four robots in the environment at the most, each one associated with a voice in JaVOX. Table 10.1 shows the relationship between the sound performance controls of JaVOX and the behavior of the mobile robots. This table describes a possible association considering the proximity between the mobile robots.

## 10.7  Conclusion

Different convex mirrors can be used to achieve an omnidirectional system. Among the possible ways of building an omnidirectional system, a spherical mirror was selected because of its availability and low cost and also because it can be used in robotic navigation and tele-operations. A significant gain in precision was obtained by applying a genetic algorithm to refine the coefficients of the perspective projection function. The use of the previous model, originally developed for a hyperbolic mirror, was very convenient. Populations of different size were tried, and the convergence was fast.

This technique does not use any specific model of the omnidirectional sensor. The resulting device is easily reproducible and of low cost. The application of the OmniEye in the AURAL environment, besides the feedback, allows to record sessions to study the behavior of the robots. In the context of arTbitrariness, the OmniEye can be considered not only as a support for creative explorations, but also as a device to learn about "automatic aesthetics". In either case, it helps the user and the computer to work together interactively in a new way to produce results that could not be produced individually.

## Acknowledgements

We wish to thank the students Lucas Soares, Igor Dias, Igor Martins and Eduardo Camargo, who worked in the development of the OmniEye. We also thank the students Thiago Spina, Felipe Augusto and Gustavo de Paula, who worked with the robots Nomad and Roomba. We thank the researchers Rubens Machado and Helio Azevedo for their useful suggestions. We are also thankful to the technical support of Jonnas Peressinotto and Douglas Figueiredo. We thank the Scientific Initiation Program of the National Research Council (PIBIC/CNPq), CenPRA and the Interdisciplinary Nucleous for Sound Studies of the State University of Campinas (NICS/UNICAMP) for making this research possible. This research work is part of the AURAL project, supported by the Foundation for the Research in São Paulo State (FAPESP) process 05/56186-9.

# References

1. Moroni, A., Von Zuben, F.J., Manzolli, J.: ArTbitration: Human-Machine Interaction in Artistic Domains. Leonardo 35(2), 185–188 (2002)
2. Moroni, A.S., Manzolli, J., Von Zuben, F.: ArTbitrating JaVox: Evolution Applied to Visual and Sound Composition. In: Ibero-American Symposium in Computer Graphics, pp. 97–108 (2006)
3. Manzolli, J., Verschure, P.F.M.J.: Roboser: a Real-world Musical Composition System. Computer Music Journal 29(3), 55–74 (2005)
4. Wassermann, K.C., Eng, K., Verschure, P.F.M.J., Manzolli, J.: Live Soundscape Composition Based on Synthetic Emotions. IEEE Multimedia 4, 82–90 (2003)
5. Murray, J., Wermter, S., Erwin, H.: Auditory robotic tracking of sound sources using hybrid cross-correlation and recurrent networks. In: Proceedings of the International Conference on Intelligent Robots and Systems, pp. 3554–3559 (2005) doi:10.1109/IROS.2005. 1545093
6. Moroni, A., Cunha, S.: OmniEye: A Spherical Omnidirectional System for Tracking Robots in the AURAL Environment. In: Proceedings of the 11th Computer Graphics and Artificial Intelligence Conference, pp. 109–118 (2008)
7. Strelow, D.W., Singh, S.: Reckless motion estimation from omnidirectional image and inertial measurements (2003), http://www.cs.wustl.edu/~pless/omnivis-Final/Strelow.pdf (accessed October 22, 2007)
8. Yagi, Y.: Omnidirectional sensing and its applications. IEICE Transactions on Information and Systems E82-D (3), 568–579 (1999)
9. Svoboda, T., Pajdla, T.: Epipolar Geometry for Central Catadioptric Cameras. International Journal of Computer Vision 49(1), 23–37 (2002)
10. Mei, C., Rives, P.: Single View Point Omnidirectional Camera Calibration from Planar Grids. In: IEEE International Conference on Robotics and Automation, pp. 3945–3950 (2007)
11. Micusik, B., Pajdla, T.: Autocalibration & 3D Reconstruction with Noncentral Catadioptric Cameras. In: IEEE Conference on Computer Vision and Pattern Recognition, pp. I-58–I-65 (2004)
12. Ying, X., Hu, Z.: Catadioptric Camera Calibration Using Geometric Invariants. IEEE Transactions on Pattern Analysis and Machine Intelligence 26(10), 1260–1271 (2004)
13. Ying, X., Hu, Z.: Can We Consider Central Catadioptric Cameras and Fisheye Cameras within a Unified Imaging Model? Eighth European Conference in Computer Vision. In: Pajdla, T., Matas, J(G.) (eds.) ECCV 2004. LNCS, vol. 3022, pp. 442–455. Springer, Heidelberg (2004)
14. Sturm, P., Ramalingam, S.: A Generic Concept for Camera Calibration. In: Pajdla, T., Matas, J(G.) (eds.) ECCV 2004. LNCS, vol. 3022, pp. 1–13. Springer, Heidelberg (2004)
15. Bakstein, H., Pajdla, T.: Panoramic mosaicing with a 180 field of view lens. In: Proceedings of the IEEE Workshop on Omnidirectional Vision, pp. 60–67 (2002)
16. Gluckman, J., Nayar, S.K.: Ego-motion and omnidirectional cameras. In: Sixth International Conference on Computer Vision. pp. 999–1005 (1998)
17. Scaramuzza, D., Martinelli, A., Siegwart, R.: A Flexible Technique for Accurate Omnidirectional Camera Calibration and Structure from Motion. In: Fourth IEEE International Conference on Computer Vision Systems ICVS 2006, pp. 45–45 (2006)
18. Scaramuzza, D.: Omnidirectional Camera Calibration Toolbox for Matlab (2008), http://asl.epfl.ch/~scaramuz/research/Davide_Scaramuzza_files/Research/OcamCalib_Tutorial.htm (accessed June 5, 2008)

19. Holland, J.H.: Adaptation in Natural and Artificial Systems. University of Michigan Press (1975)
20. Michalewicz, Z.: Genetic Algorithms + Data Structures = Evolution Programs. Springer, Heidelberg (1996)
21. Comaniciu, D., Meer, P.: Robust Analysis of Feature Spaces: Color Image Segmentation. In: IEEE Conference on Computer Vision and Pattern Recognition, pp. 750–755 (1997)
22. Moroni, A., Manzolli, J., Von Zuben, F.J., Gudwin, R.: Vox Populi: An Interactive Evolutionary System for Algorithmic Music Composition. Leonardo Music Journal 10, 49–54 (2000)
23. Moroni, A., Manzolli, J., Von Zuben, F.J., Gudwin, R.: Vox Populi: Evolutionary Computation for Music Evolution. In: Bentley, P., Corne, D. (eds.) Creative Evolutionary Systems, pp. 205–221. Morgan Kaufmann, San Francisco (2002)

# 11

# Fuzzy Logic and Genetic Algorithm Application for Multi Criteria Land Valorization in Spatial Planning

Zikrija Avdagic, Almir Karabegovic, and Mirza Ponjavic

**Abstract.** This chapter investigates the possibility of multicriterial land valorization in land use planning by the application of genetic algorithm and fuzzy logic. One of the key tools for the design of the decision support system based on this methodology is a geographic information system which serves to quantify multicriterial data and presents resulting spatial data. The methodology and the algorithm are applied to a specific problem of the spatial planning in Tuzla Canton, Bosnia and Herzegovina. In spatial multi criteria analyses, geographic information systems are used to identify alternatives, present them and give information to decision makers for evaluation, comparison and ordering of the alternatives. The limitations of multi criteria analyses in the standard GIS are necessity in defining of all the steps in advance and inability to simple change the criteria or thresholds later. Fuzzy set methodologies and GA optimization could be excellent for designing efficient tools to support the spatial decision making process. This works analizes the incorporation of these methodologies into a DBMS repository for the application domain of GIS [6]. It is shown how the useful concepts of fuzzy set theory and GA may be adopted for the representation and analysis of geographic data, whose uncertainty is an inherent characteristic.

## 11.1 Introduction

One of the key products in urban spatial planning is digital land use map, which with a system of settlements and traffic infrastructure plan, describes the spatial organization. Using the optimal model of multicriterial land use valorization with geographic information system (GIS), this map could be automatically generated . The model requests a methodology built on the existing principals of spatial planning and based on both GIS and multicriterial spatial analysis applications. The methodology could be used for the development of the decision support system in spatial planning [12].

This works presents the development of a methodology for finding the optimal model of multicriterial land valorization in land use planning by application of genetic algorithm (GA) and fuzzy logic. One of the key tools for the design of the decision support system based on this methodology is the geographic information system which serves to quantify multicriterial data and presents the resulting spatial data. The methodology and the algorithm are applied to a specific problem of the spatial planning in Tuzla Canton, Bosnia and Herzegovina.

D. Plemenos, G. Miaoulis (Eds.): Arti. Intel. Techn. for Comp. Graph., SCI 159, pp. 175–198.
springerlink.com                    © Springer-Verlag Berlin Heidelberg 2009

## 11.2   Multicriterial Optimization Problems in Spatial Planning

The optimization problem always exists when there are more alternatives in space, among which one should select the most acceptable. So, problem is related to multicriterial optimization [2].

Spatial planning methodology consists of three phases: analysis, synthesis and planning. Fig. 11.1 shows procedures during the multicriterial analysis in spatial planning.

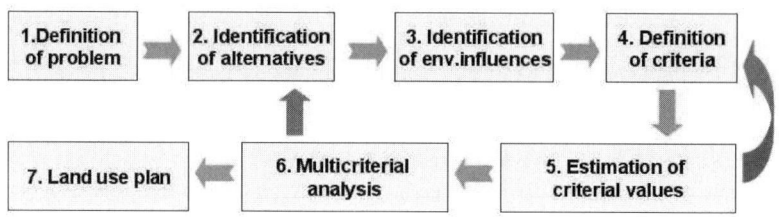

**Fig. 11.1.** Multicriterial analysis process

In the synthesis phase of planning, opposed alternatives for spatial organization are typically presented by the synthesis models. An environmental model based on protection of environment is defined due to environmental trend in urban spatial planning ,and as such, favors criteria which guarantee an environmental values continuity. Another one is functional, based on the aspect of maximum possible land exploitation for settling, not seeing the environmental consequences.

By simulating these variants by adjusting the criterial weights it is possible to search for the optimal model of spatial organization. This fact is designated by the possibility of genetic algorithm application in multicritrial optimization process.

## 11.3   Spatial Representation

Tuzla Canton, enclosing 13 municipalities with total area of 2700 km2 was used as the example in this paper. For spatial representation of the region of interest authors used 2-dimensional grid of cells (land units), arranged in rows and columns, and with the resolution of 100mx100m (Fig.11.2.). A database record, i.e. a set of attributes related to the unit properties in sense of its accessibility was added to each of land units. By these attributes, scoring and classification of the units was performed during the multicriterial analysis. The optimal model of land valorization is the result of processing data related to specific criteria [4].

GIS was used for realization of 2-dimensional grid, its integration with criterial database and geographic thematic representation of results [5]. Additionaly, this GIS approach makes planners enable to use various spatial information formats such as vector or raster data and convert them into spatial multicriterial data.

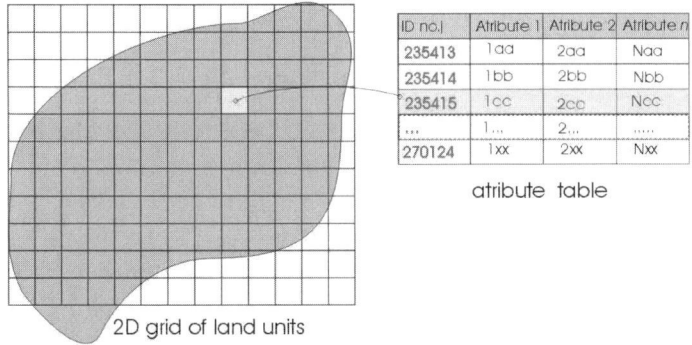

| ID no. | Atribute 1 | Atribute 2 | Atribute n |
|--------|-----------|-----------|-----------|
| 235413 | 1aa | 2aa | Naa |
| 235414 | 1bb | 2bb | Nbb |
| 235415 | 1cc | 2cc | Ncc |
| ... | 1... | 2... | ...... |
| 270124 | 1xx | 2xx | Nxx |

atribute table

2D grid of land units

**Fig. 11.2.** Two-dimensional grid of cells

## 11.4  Criterial Factors and Categories Used for Land Use Valorization

For land use valorization, the following criterial factors are used:

- Land accessibility (related to the center of settlement),
- Slope of terrain,
- Relative height (above lowest point) of terrain,
- Aspect of terrain,
- Value of land usable for agriculture and forestry (according to adopted soil classification) and
- Environmental value of vegetation coverage (estimated according to basic topographic classification and CORINE methodology).

For synthesis models (functional and environmental) are introduced the following four categories of land use:

- Extraordinary suitable,
- Very suitable,
- Suitable and
- Unsuitable.

Extraordinary suitable category is related to the area for reconstruction, and the basic use is mixed: collective and individual dwellings with central functions (e.g. services, administration etc.).

A very suitable category is related to the area of the intensive urbanization with collective and individual dwelling units, industrial and recreation zones.

A suitable one is related to the area of the extensive urbanization with mostly individual dwelling units, rural agricultural production and small business.

An unsuitable category includes two sub-categories. One is related to area mostly intended for agricultural production, and only exceptional for other uses. Another, which is related to area reserved only for forestry and agriculture, was not considered here.

## 11.5   The Application of GIS in Data Preparing for Multicriterial Analysis

The study of multicriterial land valorization of Tuzla Canton starts with the application of geographic information system. Fig.11.3 shows procedures (analytical process model) of data production for multicriterial analysis e.g. scanning plans, georeference, vectorization, polygonization, data integration etc.

Fig.11.4 shows a 3D model of Tuzla Canton that enables the classification of heights, aspects and the slopes of terrain. By the defined parameters for every class of aspect and slope, it is possible to create a thematic map in GIS.

The assignment of criterial values to classes is realized by the selection of objects belonging to a specific class and by the attachment of common attribute values in database.

Five classes are defined for these aspects: east, west, north, south and horizontal. Slopes are divided into five classes: flat, small inclination, inclined, steep and very steep. During the classification of relative heights, three zones are used for scoring: plain, hill and mountain land. There are , three categories of land bonity used to classify usability, and the classification of land accessibility is based on chronometric analysis realized in GIS. Usage of CORINE methodology provided the classification of environmental value of land [17].

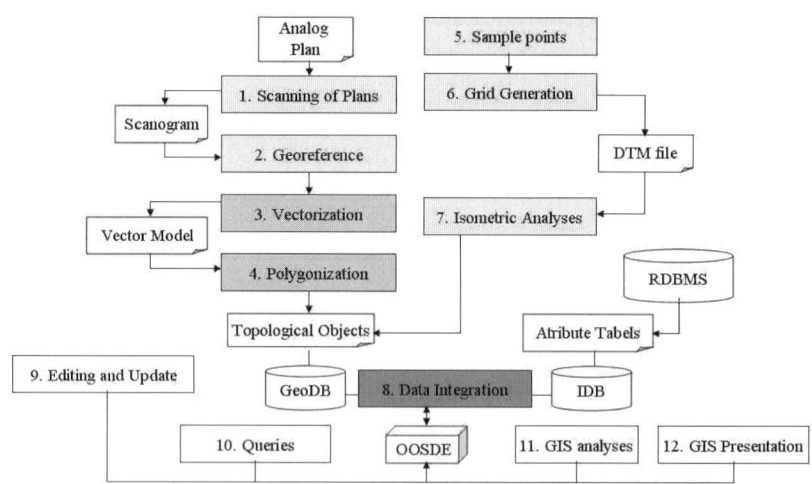

**Fig. 11.3.** Analytical Process Model of Data Production in GIS

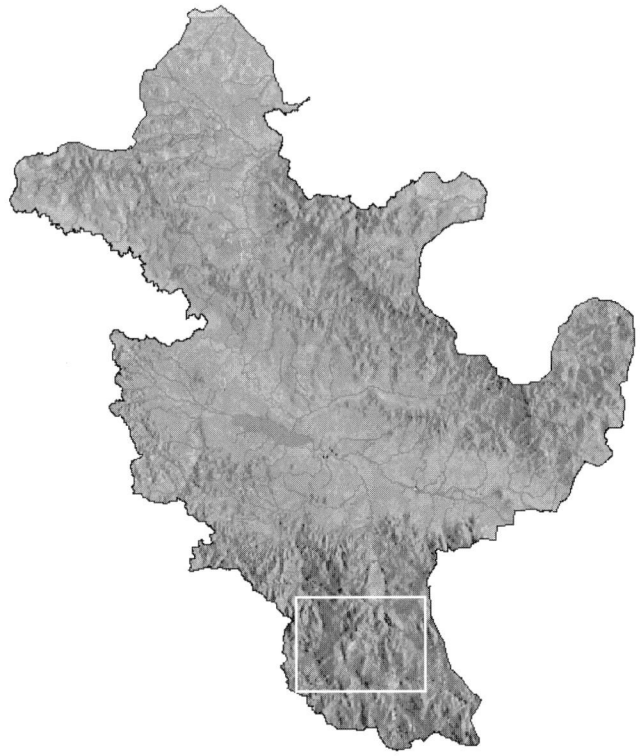

**Fig. 11.4.** 3D Model of Tuzla Canton

## 11.6  Advantages of the Fuzzy Logic Approach

GIS, at the present, has several limitations which make them an inefficient tools for decision-making. The biggest limitation is that current commercial systems are based on an inappropriate logical foundation. Current GIS are predominantly based on Boolean logic [8].

Fuzzy logic is an alternative logical foundation coming from artificial intelligence (AI) technology with several useful implications for spatial data handling. Contrary to traditional logic, fuzzy logic accommodates the imprecision in information, human cognition, perception and thought. This is more suitable for dealing with real world problems, because human reasoning is mostly imprecise.

The major advantage of this fuzzy logic theory is that it allows the natural description, in linguistic terms, of problems that should be solved rather than in the terms of relationships between precise numerical values. This advantage, dealing with the complex systems in simple way, is the main reason why fuzzy logic theory is widely applied in technique.

Fuzzy logic appears to be instrument in the design of efficient tools for spatial decision making. Fuzzy set theory is an extension of the classical set theory. A fuzzy set A is defined mathematically as follows:

IF X = {x} denotes a space of objects, THEN the fuzzy set A in X is the set of ordered pairs: A = {x, μA(x)}, x ∈ X,

where the membership function μA(x) is known as the "degree of membership (d.o.m.) of x in A". Usually, μA(x) is a real number in the range [0, 1], where 0 indicates no-membership and 1 indicates full membership. Here μA(x) of x in A specifies the extent to which x can be regarded as belonging to set A.

The operations of fuzzy set theory provide the counterpart operations to those of classical set theory. Logical operations with fuzzy sets are more generalized forms of usual Boolean algebra applied to observations that have partial membership of more than one set. The standard operations of union, intersection, and complement of fuzzy sets A and B, defined in domain X, create a new fuzzy set whose membership function is defined as:

$$\text{Union:} \qquad \mu A\,B(x) = \max\{\mu A(x), \mu B(x)\}, x \in X \qquad (11.1)$$

$$\text{Intersection:} \quad \mu A\,B(x) = \min\{\mu A(x), \mu B(x)\}, x \in X \qquad (11.2)$$

$$\text{Complement:}\ \mu{\sim}A(x) = 1 - \mu A(x), \qquad\qquad x \in X \ (11.3)$$

For example, let us consider the classification of individual locations on a layer based on the slope values with linguistic values [level, gentle, moderate, steep] and a second classification based on the land moisture with linguistic values: [dry, moderate, wet, water].

For each individual location l (e.g., d.o.m. for level = 0.8 and d.o.m. for dry = 0.4) the d.o.m. value which provides an overall measure regarding:

1. level ground and dry land is derived by: $\min\{\mu level(l), \mu dry(l)\}$, (e.g., $\min\{0.8, 0.4\} = 0.4$);
2. level ground or dry land is derived by: $\max\{\mu level(l), \mu dry(l)\}$, (e.g., $\max\{0.8, 0.4\} = 0.8$); and
3. non-level ground is derived by: $1 - \mu level(l)$, (e.g., $1 - 0.8 = 0.2$).

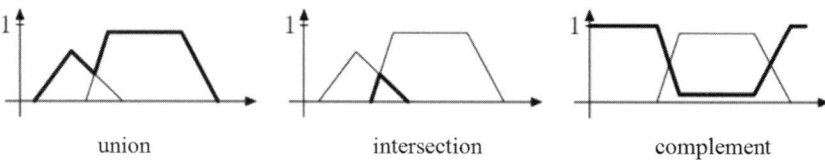

union                    intersection                    complement

**Fig. 11.5.** Standard fuzzy operations

## 11.7  Limitations of GIS

Uncertainty in GIS means the imperfect and inexact information. The uncertainty is an inherent feature of geographic data. Currently used methods for the representation and analysis of geographic information are inadequate, because they do not tolerate uncertainty. This is mostly due to the applied membership concept of the classical set theory, where a set has precisely defined boundaries and an element has either full or no membership in the set (Boolean logic).

The representation of geographic data based on the classical set theory affects reasoning and analysis procedures, adding in all the problems of an "early and precisely classification". The final decision is made after steps which drastically reduced the intermediate results [9]. Any constraint is accompanied with an absolute threshold value and no exception is allowed. For instance, if the threshold for a flat land is slope = 10%, a location with slope equal to 9.9% is characterized as level, while a second location with slope equal to 10.1% is characterized as non-level (steep). Moreover, for decisions based on multiple criteria, it is usually the case that an entity (i.e., an individual location), which satisfies quite well the majority of constraints and is marginally rejected in one of them, to be selected as valid by decision-makers.

However, based on Boolean logic, a location with slope 10.1% will be rejected (as non-level), even if it satisfies quite well all other constraints posed by decision-makers. In addition, decision-makers are obliged to express their constraints through arithmetical terms and mathematical symbols in crisp relationships (e.g., slope < 10%), since they are not allowed to use natural language linguistic terms (e.g., flat land). Finally, another effect of classical set theory is that the selection result is flat, in the sense that there is no overall ordering of the valid entities as regard to the degree they fulfill the set of constraints. For instance, dry-level layer highlights all locations which satisfy the constraints: dry land (threshold 20%) and flat land (threshold 10%). However, there is no clear distinction between a location with moisture = 10% and slope = 3% and another with moisture = 15% and slope = 7%. These impediments call for a more general and sound logical foundation for GIS.

### 11.7.1 Overlay Operation

The polygon overlay is used to calculate the common (or different) area between two overlapping objects.

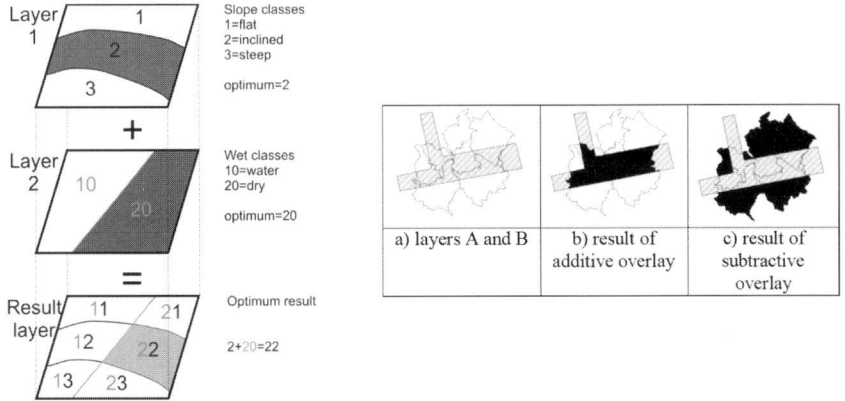

**Fig. 11.6.** Schema of polygon overlay operation

The overlay operation is analogous to join operation in conventional database systems, and is defined as the assignment of new attribute values to individual locations resulting from the combination of two or more layers [10].

It could be additive or subtractive overlay operation. The additive overlay creates one or more polygons from the intersection between the polygons on layers A and B. The subtractive overlay subtracts the polygons on layer 2 from the polygons on layer 1. Fig. 11.6. a) shows two layers: layer A (no hatch pattern) and layer B (diagonal hatch). The result of additive overlay on layer C is displayed in solid black on b). The result of a subtractive overlay on layer C is displayed in solid black on c). Overlay operation is most used of all vector analysis in GIS [14].

### 11.7.2   Site Selection in Classic GIS

This section will present an example of searching for relevant location, study done as a part of making The Spatial Plan of Tuzla Canton. In this situation, the set of constraints and opportunities consists of: level and smooth site (slope < 20%), not-north-facing slope, not agriculture land (usability) class 1, not close to garbage depot, vacant area (no development), not close to exploiting area, not close to sliding-land area, not reserved for special purpose, not close to mine contaminated area (MCA), closeness to the existing road network, closeness to the existing electrical network, dry land.

In addition, all candidate sites should have an adequate size to satisfy the needs of the planning activity (more then 2 sq km). The whole task, as input, requires five themes (layers) of the region under examination: hypsography theme (3D surface of the region or altitude values), development theme (existing infrastructure of the region like roads or buildings), vegetation theme (area covered with vegetation like forest or usability areas), moisture theme (soil moisture of the region like lakes, wetlands, dry-lands) and MCA theme.

The procedure of site selection, based on the sets of constraints and opportunities determined above, may consist of the sequence of operations. First, from these 5 themes it is necessary to extrude 12 layers one by one. Some of the layers should be buffered (roads are usually presented as lines, buffer operation will make areas of them; MCA or asliding area are very dangerous, so it should be buffered to wider protective band). After that, usage of the overlay operation (additive and subtractive)

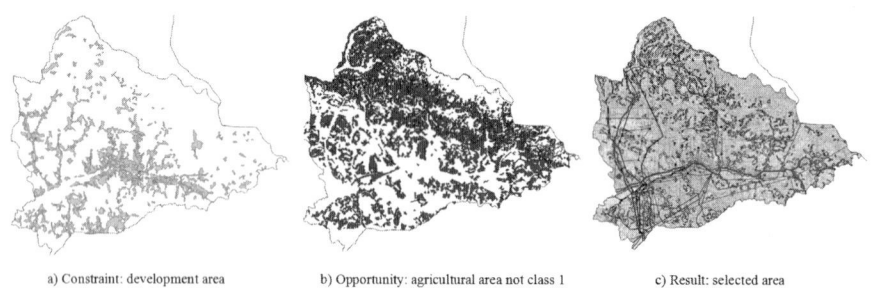

a) Constraint: development area          b) Opportunity: agricultural area not class 1          c) Result: selected area

**Fig. 11.7.** Examples of constraints and opportunities

of all layers will produce a result layer with only areas that satisfy all criteria. Then, it should be checked if candidate sites satisfy condition of minimal area and exclude which not. Two set were created: a set of constraints (e.g. development area on Fig. 11.7.a), which restrict the planned activity, and a set of opportunities (e.g. agricultural area on Fig. 11.7.b), which are suitable for the activity. The combination of these two is considered in order to find the best locations (result on Fig. 11.7.c).

This procedure was very demanding in time and it was unsuitable for decision making in real time. It produced useful results but it also emphasized some of the limitations. The biggest problem was that all criteria have to be given in advance and every change requires repeating many steps of the procedure (time demanding). Second problem was the mathematical precision of data, which is in such real case unnecessary high and requires additional user's effort to define precise constraints.

## 11.8   Fuzzy Logic in Spatial Multiple Criteria Decision Making

In a classic model, each theme is described through a set of attribute values and each individual location on it is assigned by only one of these values. The assignment of an attribute value to an individual location indicates its full membership regarding this feature in the corresponding layer.

In the fuzzy set theory the concept of full membership is replaced by the concept of partial membership and consequently the representation of individual locations should change. The incorporation of fuzziness into the spatial data model forces the redefinition of the components forming the hierarchical data model.

Specifically, while in the conventional set theory the individual locations in a layer are assigned the attribute values (soil, grass, fruit-trees, forest) characterizing a theme (vegetation), in fuzzy set theory they are assigned d.o.m. values regarding each

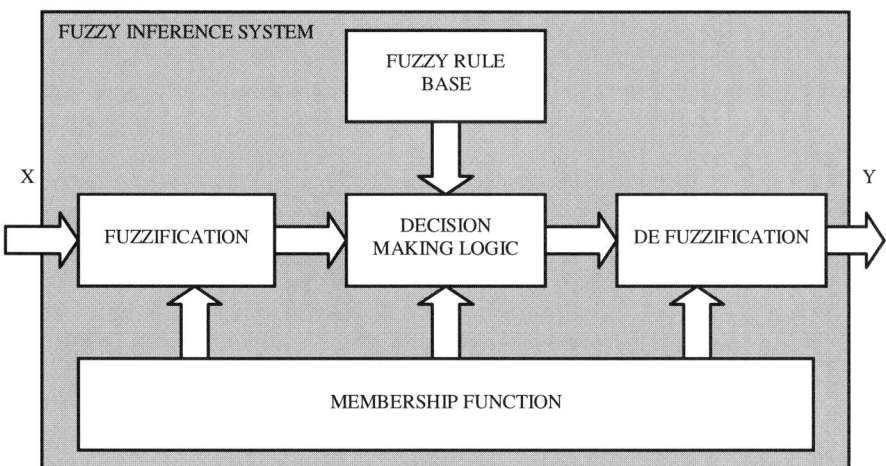

**Fig. 11.8.** Block structure of fuzzy system

attribute value (0.1 for soil, 0.6 for grass, 0.3 for fruit-trees and 0 for forest). These values are derived by applying both the appropriate membership functions chosen by decision-makers and the knowledge provided by the experts [11].

Field measurements and results derived from sampling techniques are processed and transformed into d.o.m. values for the predefined attribute (linguistic) values characterizing a theme. Apparently, the number of layers increases, since each theme is represented by as many layers as the number of attribute values associated to it.

The Fuzzy Logic Toolbox in MATLAB provides tools for building Fuzzy Inference System (FIS), as show on Fig. 11.8. Fuzzy inference is the process of formulating the mapping from a given input to an output using fuzzy logic. The process of fuzzy inference involves: membership functions, fuzzy logic operators and if-then rules. There are two types of fuzzy inference systems that can be implemented in the Fuzzy Logic Toolbox: Mamdani-type and Sugeno-type.

Mamdani's fuzzy inference method is the most commonly seen fuzzy methodology and it expects the output membership functions to be fuzzy sets. After the aggregation process, there is a fuzzy set for each output variable that needs defuzzification.

There are five parts of the fuzzy inference process: the fuzzification of the input variables, choosing membership functions, constructing rules, making decision and defuzzification.

### 11.8.1  Fuzzification

6 main criteria were chosen for the analyses: spatial accessibility (from centers of settlements, with consider of natural barriers), slope (level or slope), relative altitudes, aspect (orientation to the sun), usability (for forestry and agriculture) and ecological value (land use got from satellite images).

Also there are some constraints like areas under water (lakes and bigger rivers), sliding-land areas, forest, mining areas, construction areas, MCA etc.

An important issue for decision making is reasoning based on linguistic values assigned to physical entities (e.g. inclined is slope between 4% and 10%). A set of linguistic values should be assumed to classify entities and measurements in categories. Each linguistic value corresponds to a range of physical values. Every input criterion should be fuzzified. For example, slopes are classified in five categories and that is shown in Table 11.1.

Based on this classification, it is made a thematic map of slopes in GIS shown on Figure 11.9.

**Table 11.1.** Fuzzification of slopes

| Classes of slopes | Klase nagiba | from | to |
|---|---|---|---|
| flat (level) | ravno | 0 | 2 |
| small inclination (gentle) | mali nagib | 2 | 4 |
| inclined (moderate) | nagib | 4 | 10 |
| steep | strmo | 10 | 20 |
| very steep | vrlo strmo | 20 | 30 |

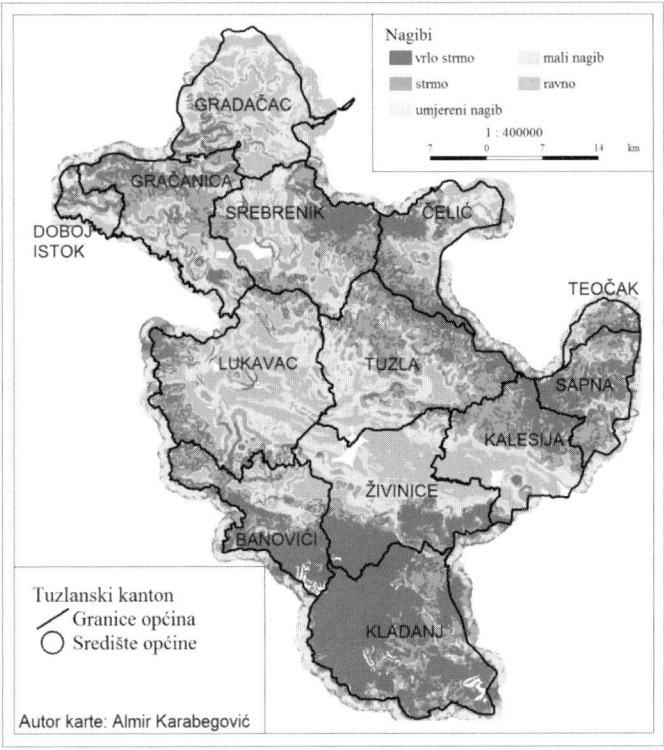

**Fig. 11.9.** Thematic map of slopes

### 11.8.2   Choosing Membership Functions

A fuzzy membership function is a curve that defines how each point in the input space is mapped to a membership value (or degree of membership) between 0 and 1. The input space is sometimes referred to as the universe of discourse. The choice of the membership function, its shape and form, is crucial and strongly affects the results derived by the decision-making process.

There are several membership function mostly used for geographical phenomena, but especially triangular and Gaussian. A Gaussian membership function is built on the Gaussian distribution curve and defined as the following formula.

$$f(x_1) = \frac{1}{\sigma\sqrt{2\pi}} e^{\frac{-1}{2\sigma^2}(x-\mu)^2} \qquad (11.4)$$

where $\mu$ is the mean and $\sigma$ is the standard deviation, the two parameters for the Gaussian membership function.

Because of its smoothness and concise notation, Gaussian membership function is a popular method for specifying fuzzy sets. This curve has the advantage of being smooth and nonzero at all points [18].

In this work, Gaussian membership function is a form used in most of criteria, as it is shown for slopes on Fig. 11.10. There is one transformation function associated to each linguistic value, what means that number of functions is equal to the number of linguistic values assumed.

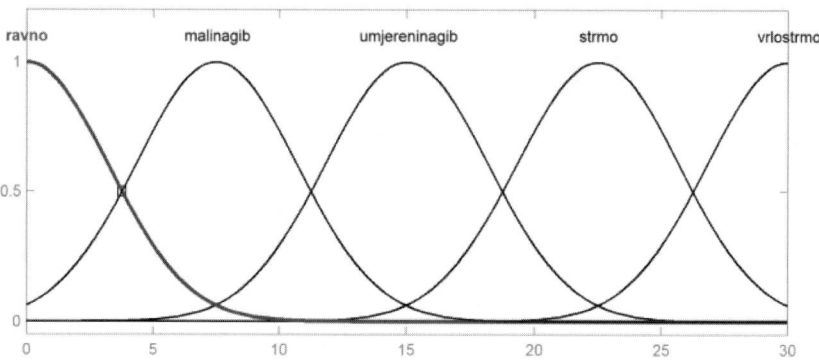

**Fig. 11.10.** Membership function for slopes

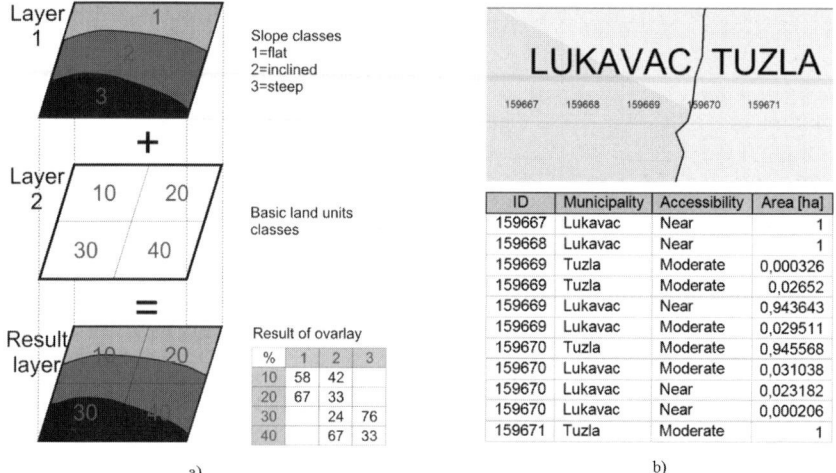

**Fig. 11.11.** Overlaying; a) schematically and b) example

### 11.8.3   The Representation of Fuzzified Values

A general spatial data model is here presented in space as a two-dimensional grid of cells, or land units. This grid is created in GIS, every cell is one entity connected with one record in the database. Schematically, it is shown in Fig.11.11. a.

Most important feature of grid is its resolution, because accuracy of results is dependent on it. In isometric analysis a grid resolution of 30x30 m2 is used for every municipality. But, for the whole Tuzla Canton area (around 2700 km2), as optimal

| Basic information | ID | Number | Municipality | |
|---|---|---|---|---|
| | 1000554913 | 188342 | Tuzla | |

| Slope classes | Flat | Small inclination | Inclined | Steep | Very steep |
|---|---|---|---|---|---|
| | 0,036906179 | 0,251406435 | 0,479215376 | 0,211899903 | 0,020572108 |

| Aspect classes | Horizontal | North | East | South | West |
|---|---|---|---|---|---|
| | | | 0,71126 | 0,28874 | |

| Accessability classes | Close | Near | Moderate far | Far | Very far |
|---|---|---|---|---|---|
| | | | | 1 | |

| Altitudes classes | Lawland | Hill | Mountain |
|---|---|---|---|
| | | 1 | |

| Agriculture classes | Agrozone1 | Agrozone2 | Agrozone3 |
|---|---|---|---|
| | | 0,0333171 | |

| Biological values classes | NoBioVal | SmallBioVal | MedBioVal | HighBioVal | VeryHighBioVal |
|---|---|---|---|---|---|
| | 0,30296106 | | 0,69703894 | | |

| Development classes and wet | Development | Economic | Water |
|---|---|---|---|
| | 0,9415659 | | |

**Fig. 11.12.** Table of fuzzified values

resolution, grid 100x100 m, or 1 ha, is chosen and that satisfies the level of regional planning. This produced the layer of basic land units (grid layer) with 281,526 entities and the same number of records in database.

This layer is overlaid with every theme (layer). This operation adds linguistic values (classes) of overlaid layer as new attributes to the table of basic land units, and fills in values such as a size of area that covers that class in that cell. An example is overlaying the layer of accessibility with the layer of basic land units. Results are shown in Fig.11.11. b. As a result there is a big table connected to the layer of basic land units with all classes of all input criteria as attributes and with their parts in area as values. This table of fuzzified values for one record is presented in Fig.11.12

### 11.8.4 Constructing Rules

Fuzzy sets and fuzzy operators are the subjects and verbs of fuzzy logic. These if-then rule statements are used to formulate the conditional statements that comprise fuzzy logic. A single fuzzy if-then rule assumes the form:

$$\text{IF } x \text{ IS } A \text{ THEN } y \text{ IS } B \tag{11.5}$$

where A and B are linguistic values defined by fuzzy sets on the ranges (universes of discourse) X and Y, respectively. The if-part of the rule "x is A" is called the antecedent or premise, while the then-part of the rule "y is B" is called the consequent or conclusion. An example of such a rule might be

$$\text{IF slope IS inclined THEN area IS suitable} \tag{11.6}$$

The input to an if-then rule is the current value for the input variable (slope) and the output is an entire fuzzy set (suitable). This set will later be defuzzified, assigning one value to the output.

Interpreting an if-then rule involves distinct parts: first evaluating the antecedent (which involves fuzzifying the input and applying any necessary fuzzy operators) and second applying that result to the consequent (known as implication). In the case of two-valued or binary logic, if-then rules don't present much difficulty. If the premise is true, then the conclusion is true. If the antecedent is true to some degree of membership, then the consequent is also true to that same degree. The antecedent of a rule can have multiple parts.

> IF (slope IS flat) AND (aspect IS south) AND (accessibility IS close) AND (altitudes IS low) AND (usability IS agrozona3) THEN area IS suitable (1)

in which case all parts of the antecedent are calculated simultaneously and resolved to a single number using the logical operators. The number in the brackets is the weight of that rule. Every rule has a weight (a number between 0 and 1), which is applied to the number given by the antecedent. Generally this weight is 1 and so it has no effect at all on the implication process.

Apparently, the number of rules is equal to the number of combinations for all membership functions (classes, linguistic values). Using knowledge base, some of combinations are excluded, so final number of rules for solving this problem is 432.

The consequent specifies a fuzzy set that will be assigned to the output. After that, the implication function modifies that fuzzy set to the degree specified by the antecedent. The most common way to modify the output fuzzy set is truncation using the min function.

Consider the rule (11.5). If it is observed that x is A', it uses fuzzy implication to reason that y is B'. Mathematically written, the implication form is

$$R = \int_{(x,y)} \mu(x,y)/(x,y) \quad \text{or} \quad R = \sum_{(x_i,y_i)} \mu(x_i,y_i)/(x_i,y_i) \tag{11.7}$$

There are 40 implication operators, but most important are Zadeh Max-Min, Mamdani Min and Larsen. Mamdani Min implication operator is used , defined as:

$$\Phi m\,[\mu A(x), \mu B(y)] \equiv \mu A(x) \wedge \mu B(y) \rightarrow \mu(x, y) \tag{11.8}$$

where $\Phi$ is an implication operator which takes as an input membership function of antecedent $\mu A(x)$ and consequent $\mu B(y)$.

### 11.8.5  Decision Making

Fuzzy algorithms are evaluated using generalized modus ponens (GMP). GMP is a data-driven inferencing procedure that analytically involves the composition of fuzzy relations, usually max-min composition. Max-min composition under a given implication operator affects the right side of the rule in a specific manner (by clipping with Mamdani or scaling with Larsen implication operator). In general, GMP is a transformation of the right side of the rule by a degree commensurate with the degree of fulfillment (DOF) of the rule and in a manner dictated by the chosen implication operator. As far as the entire algorithm is concerned, the connective ELSE is analytically modeled as either OR ($\vee$) or AND ($\wedge$), depending on the used implication operator for the individual if-then rules (when the Mamdani min implication is used, the connective ELSE is interpreted as OR). In this work is used GMP with many inputs and many rules.

From the table of fuzzified values, using given rules, it is now possible to make multiple criteria analysis or multiple criteria decision making [7]. The easiest way of manipulation data in tables is with SQL statements. So, in this work it is suggested to transfer if-then rules to SQL statement. Last if-then rule could be present in database as:

```
SELECT
ID, Municipality
FROM
TK
WHERE
Slope Is Not Null AND South Is Not Null AND Close Is Not Null AND Low Is
Not Null AND [Agrozona 3] Is Not Null;
```

All basic land units which satisfy this condition and calculated total area are selected based on such query in GIS. The result is same as the one derived inform the classical method of overlaying and there no any ranging of data [19].

A problem that arises in this case is that only one of the participating d.o.m. values dominates by assigning its value to the whole decision criterion. In this way the contribution of the other d.o.m. values is eliminated.

For decision criteria which combine more than one layer and linguistic value e.g. level ground and dry land an overall measure should be computed and assigned to individual locations. This measure is derived from the consideration of d.o.m. on two or more layers. For a fuzzy set $A \in X$ with d.o.m. $\mu A(x) \in [X]$ the overall measure can be provided by an exponential function, which is given by the following commonly used formula:

$$\mu E(x) = \sum_{i=1}^{k} [\mu_{A_i}(x)]^q \tag{11.9}$$

By applying this equation (e.g. for q = 2, quadratic measure) the big weight values (d.o.m.) are amplified, while the small values are nearly eliminated. Assuming the previous example, the overall measure characterizing each individual location (l) of a region, regarding level ground and dry land using the energy function, is given by:

$$\mu_{level-dry}(l) = [\mu_{level}(l)]^2 + [\mu_{dry}(l)]^2 \tag{11.10}$$

Results derived by the previous formula should be normalized in the fuzzy domain [0, 1]. Using formula (11.10) produced a new SQL statement which added a new result field to the express degree of membership of every basic land unit.

```
SELECT
ID, Municipality, ([Flat]^2 + [South]^2 + [Close]^2 + [Low]^2 + [Agrozone 3]^2)
AS Result
FROM
TK
WHERE
Flat Is Not Null AND South Is Not Null AND Close Is Not Null AND Low Is Not
Null AND [Agrozone 3] Is Not Null;
```

Here exponent 2 is chosen and it provides the order of qualified locations. This feature of exponent is very beneficial for decision criteria which combine multiple sets and linguistic values and make order of results for decision maker.

In GIS, the process of visualization of such data is the process of making thematic maps, which gives a decision maker a clear picture of his decision [16].

### 11.8.6  Defuzzification

The output of fuzzy system is a fuzzy value. There is an option of using this value without any modification (leaving the final crisp action to the human operator) or to use a defuzzification scheme and generate a crisp output. Commonly used defuzzification schemes include Tsukamoto's, Center of Area (COA) and Mean of Maximum (MOM) methods.

Fuzzy output was defined in four classes as in Table 11.2. These linguistic values are from the  real world, and there are terms that the decision makers normally use in their work. So, even without any modification (leaving fuzzy values) results are appropriate.

**Table 11.2.** Fuzzification output

| Category classes | Klase kategorizacije | from | to |
|---|---|---|---|
| extraordinarily suitable | izvanredno podobna | 75 | 100 |
| very suitable | vrlo podobna | 50 | 75 |
| suitable | podobna | 25 | 50 |
| unsuitable | nepodobna | 0 | 25 |

### 11.8.7  Fuzzy Query

Final query for multiple criteria decision making did land valorization for every of 13 municipalities in Tuzla Canton, and the final result for Tuzla municipality is in Fig. 11.13. It is also produced a table of areas balances for all municipalities, where decision makers can see, for every class of quality, how big is the area it covers. Combined with thematic maps, it makes the base for any analyses.

Getting results with such procedures is only the matter of database and GIS is now just a tool for making the spatial presentation of the results. Contrary to the classic method, where everything was done graphically (in database data were only copied), this methodology employs database, which enables us to put the time demand part in preparation and defining criteria delaying to time of creating queries.

Every change of input data, now, requires only checking its influents to information (classic UPDATE statement in database). Also, data is ordered according to its importance for decision makers.

```
SELECT
ID,
Municipality,
[flat]+[small_inclination]+[inclined] AS Slope,
[east]+[south]+[west] AS Aspect,
[close]+[near]+[moderate_far] AS Accessibility,
[lawland]+[hill] AS Altitude,
[agrozone2]+[agrozone3] AS Usability,
[nobioval]+[smallbioval]+[medbioval] AS BioValue,
([slope]^2+[aspect]^2+[accessibility]^2+[altitude]^2+
[usability]^2+[biological_value]^2)/6 AS Result
FROM
TK
WHERE
[flat]+[small_inclination]+[inclined])>0) AND
(([east]+[south]+[west])>0) AND
(([close]+[near]+[moderate_far])>0) AND
(([lawland]+[hill])>0) AND
(([agrozone2]+[agrozone3])>0) AND
(([nobioval]+[smallbioval]+[medbioval])>0) AND
((ALL CONSTRAINTS)=0));
```

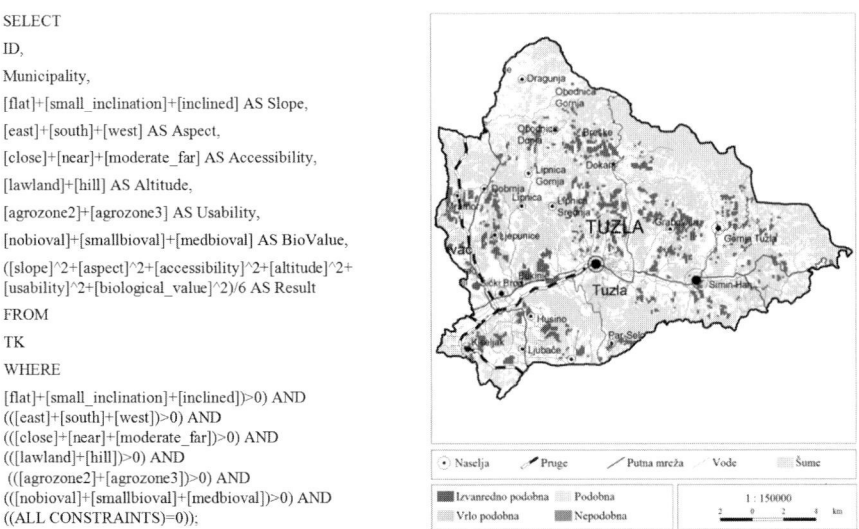

**Fig. 11.13.** Land valorization for municipality Tuzla, SQL query and graphic result

## 11.9   Multicriterial Classification and Assigning of Fuzzy Values

All classes specified above are scored in a scope from 1 to 5 points (Fig. 11.14).

| Slope of Terrain Classes | Slope of Terrain Description | Aspects of Terrain Classes | Aspects of Terrain Description | Relative Hights Classes | Relative Hights Description | Land Usable Value Classes | Land Usable Value Description | Environmental Land Value Classes | Environmental Land Value Description | Land Accessibility Classes | Land Accessibility Description | SCORES (1-5) |
|---|---|---|---|---|---|---|---|---|---|---|---|---|
| flat | 0-2% | horiz. | 0-360° | plain land | 0-300m | 1st agrozone | I-IVa category | very low value | 2.3.1, 3.2.1 | very near | 0-5min | 5 |
| small inclinat. | 2-4% | South | 135-225° | - | - | - | - | low value | 2.2.1 | near | 5-10min | 4 |
| inclined | 4-10% | East/ West | 45-135°/ 225-315° | hill land | 300-700m | 2nd agrozone | IVb-VI category | middle value | 2.2.2, 2.4.3, 3.2.2 | accesible far | 10-15min | 3 |
| steep | 10-20% | - | - | - | - | - | - | high value | 2.4.1, 2.4.4, 3.2.3, | far | 15-20min | 2 |
| very steep | 20-30% | North | 0-360° | mountain land | above 700m | 3rd agrozone | VII-VIII category | very high value | 3.1.1, 3.1.2, 3.1.3 | very far | 20-30min | 1 |

**Fig. 11.14.** Table of scoring of land units

Each of land units was assigned a fuzzy value, related to the unit properties. By these values, scoring and classification of the unitsis performed during the multicriterial analysis. The optimal model of land valorization is the result of processing of data related to specific criteria [3].

Additionally, the land for reconstruction is determined by the analysis of existing construction areas, as it is shown in Table. 11.3. These areas will be used to create fitness function for multicriterial optimization.

**Table 11.3.** Balance of areas used for reconstruction

| Order Number (k) | Municipality | Area Used for Reconstruction ($P_{ok}$) in hectares |
|---|---|---|
| 1 | Banovici | 165 |
| 2 | Celic | 341 |
| 3 | Doboj Istok | 383 |
| 4 | Kalesija | 124 |
| 5 | Gracanica | 422 |
| 6 | Gradacac | 770 |
| 7 | Kladanj | 578 |
| 8 | Lukavac | 1485 |
| 9 | Srebrenik | 155 |
| 10 | Sapna | 280 |
| 11 | Teocak | 402 |
| 12 | Tuzla | 508 |
| 13 | Zivinice | 127 |

## 11.10 The Representation of Alternatives by Synthesis Models

The weights of criteria are defined according to differences in their importance which are adopted as linear dependent values in a model. Fig 11.15 shows normalized weights for two syntesis models according to their importance related to the specific model.

According to Fig. 11.16, behavior of the model, in functional and environmental sense, is possible to be described by the following set of linear equations:

$$p_1 = -2,5 \ tg \ \alpha + 3,5$$

$$p_2 = -1,5 \ tg \ \alpha + 3,5$$

$$p_3 = -0,5 \ tg \ \alpha + 3,5$$

$$p_4 = \ 0,5 \ tg \ \alpha + 3,5$$

$$p_5 = \ 1,5 \ tg \ \alpha + 3,5$$

$$p_6 = \ 2,5 \ tg \ \alpha + 3,5 \qquad (11.11)$$

| Normalized Weights for | Land Accessibility | Slope of Terrain | Relative Height | Environmental Value | Aspect of Terrain | Land Usable Value |
|---|---|---|---|---|---|---|
| Environmental Model | 0.28 | 0.24 | 0.19 | 0.14 | 0.10 | 0.05 |
| Functional Model | 0.05 | 0.10 | 0.14 | 0.19 | 0.24 | 0.28 |

**Fig. 11.15.** Weights of criterial factors for environmental and functional model

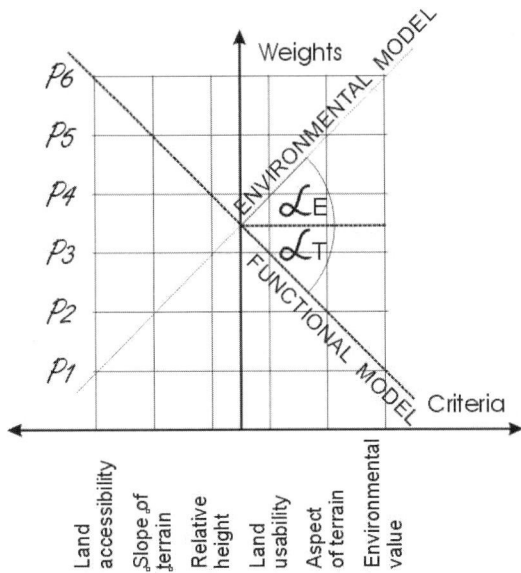

**Fig. 11.16.** Synthesis models and values of criterial weights

where $p_1$, $p_2$ ... $p_6$ denote the weights of specific criteria, and $\alpha$ is the angle of model gravitation that represents how much the model gravitates to some of the alternatives (synthesis models).

## 11.11  The Optimization of the Model

If the coefficient of direction, tang $\alpha_i$, varies from -1 to 1, then the angle of gravitation, $\alpha_i$, takes values from $-\pi/4$ to $\pi/4$. The optimum model is characterized by total suitable area P depending on the angle $\alpha$.

If existing land for construction is adopted as extraordinary suitable category, then the condition for the optimization of the model can be described by the expression:

$$F = (P_r - P_O)^2 = \min. \tag{11.12}$$

where Pr is extraordinary suitable area for optimum model, and $P_O$ is existing land for construction.

In order to make the description of the problem easier, function $P_r(\alpha)$ is presented by appropriate polynom $\psi_r$, and expression (11.12) can be transformed into:

$$\Phi = [\psi_r(\alpha) - P_O]^2 = \min. \tag{11.13}$$

which represents objective function for the optimizationof the model .

A method used for solving optimization problem is the genethic algorithm that is based on natural selection, the process that drives biological evolution. It can be applied to solve various optimization problems in which the objective function is discontinuous (or with discrete values), nondifferentible, stochastic or nonlinear.

The building blocks of the genetic algorithm are the evaluation of fitness, selection, recombination and the population of chromosomes [13].

The genetic algorithm repeatedly modifies a population of individual solutions (chromosomes). At each step, the genetic algorithm randomly selects individuals from the current population to be parents and uses them to produce the children for the next generation. Over successive generations, the population evolves toward an optimal solution [1].

### 11.11.1  Fitness Function

According to expression (10.13) fitness function $F_f$ can be described as:

$$F_f = \sum_{k=1}^{n} \Phi_k = \sum_{k=1}^{n} [\psi_{rk}(\alpha) - P_{Ok}]^2 \tag{11.14}$$

where $n$ is total number of enclosed municipalities.

The expression (11.14) is used for evaluation of fitness values necessary for the creation of each next generation of potential solutions (chromosomes).

Fitness function is defined in M-file (Matlab) by calculated polynomial coefficients and balance of areas from Table 11.3.

### 11.11.2  The Representation of Chromosome

The binary string is used fot the representation of chromosome [15]. A variable is encoded so that it presents the real values of the angle of gravitation in radians. The domain of searching is defined with adopted precision of 0.01 radian. In this way, the solutions are presented by chromosomes (104 bits strings) consisting of 13 genes.

Each of the gene (8 bits string) represents a model (by angle of gravitation) of specific municipality.

### 11.11.3  The Parameterization of GA and Results

After the testing, the parameters which gave acceptable results of optimization are determined as:  roulette wheel selection, 100 chromosomes in population, elite count 2, crossover fraction 0.25, mutation with Gaussian distribution, single point crossover and stopping after 500 generation.

Final value of fitness obtained in the last generation is 0.24 ha, while predefined value of tolerance is 2 ha (Fig. 11.17).

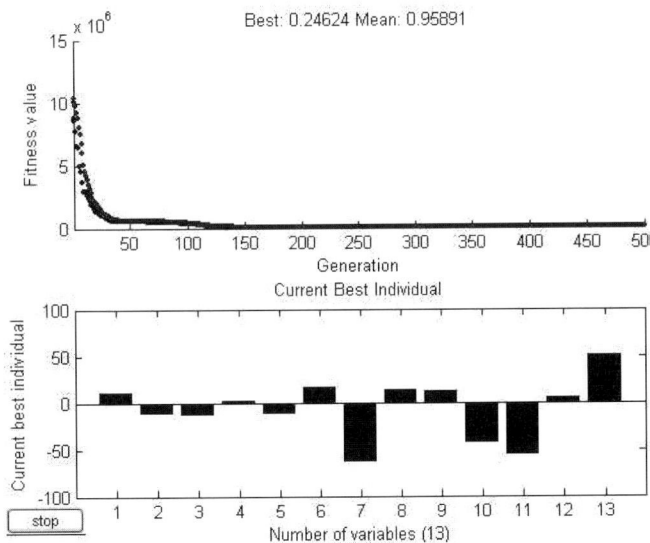

**Fig. 11.17.** Current best individual and fitness value

**Table 11.4.** Angles of model gravitation

| Order Number (k) | Municipality | MGA ($\alpha_k$) in *rad* |
|---|---|---|
| 1 | Banovici | 0.103 |
| 2 | Celic | -0.118 |
| 3 | Doboj Istok | -0.131 |
| 4 | Kalesija | 0.026 |
| 5 | Gracanica | -0.127 |
| 6 | Gradacac | 0.162 |
| 7 | Kladanj | -0.639 |
| 8 | Lukavac | 0.133 |
| 9 | Srebrenik | 0.118 |
| 10 | Sapna | -0.437 |
| 11 | Teocak | -0.571 |
| 12 | Tuzla | 0.049 |
| 13 | Zivinice | 0.507 |

Angles of model gravitation (MGA) obtained by GA optimization are given in Table 11.4.

## 11.12  Land Use Classification and Thematic Presentation in GIS

Applying the genetic algorithm for searching the optimum model for land use classification are determined values of the angles of gravitation close enough to the optimum. In order to achieve the final objective of multicriterial analysis, it is necessary to perform aggregation, i.e. summing the factorized criterial values and classifying the areas according to the land use (already described).

Total value of a land unit is calculated as:

$$\upsilon_{zk} = (w_{1k} f_1 + w_{2k} f_2 + w_{3k} f_3 - w_{4k} f_4 - w_{5k} f_5 - w_{6k} f_6 + \upsilon_{max})/(\upsilon_{max} - \upsilon_{min}) \quad (10.15)$$

where:

- $w_{ik}$ are normalized weights from equations (11.11), for criteria i=1...6 and municipalities k=1...13,
- $f_i$ is assigned scores for specific criteria, and
- $\upsilon_{max}$ i $\upsilon_{min}$ are maximal and minimal value of land unit

Table 11.5 shows ranked values used for classification of land use.

**Table 11.5.** Ranked values for land use classification

| Category of Area | Ranked Values (normalized) |
|---|---|
| extraordinary suitable | 0.75 - 1.00 |
| very suitable | 0.50 - 0.75 |
| Suitable | 0.25 - 0.50 |
| Unsuitable | 0.00 - 0.25 |

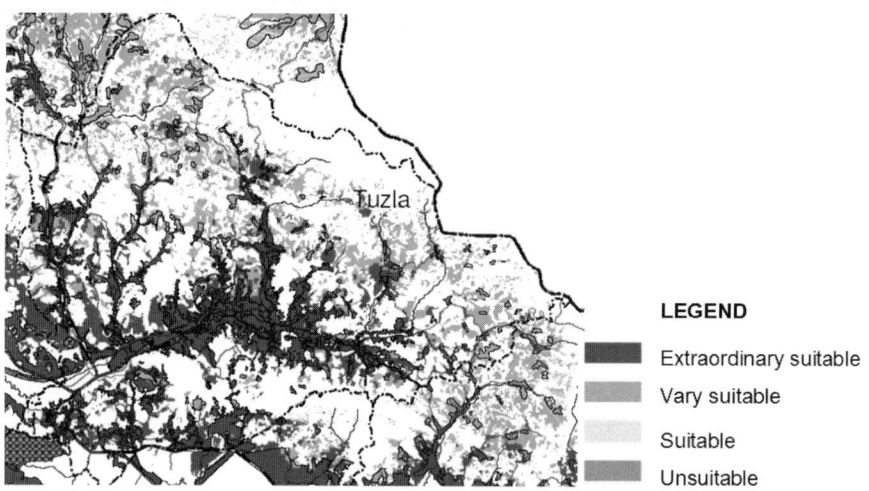

LEGEND

Extraordinary suitable

Vary suitable

Suitable

Unsuitable

**Fig. 11.18.** Thematic map of land use categories

Based on the ranked values, total land units values are presented by thematic visualization in GIS. To each of the ranks (classes) is assigned a corresponding color (Fig.11.18).

## 11.13  Conclusion

The representation of geographic data based on the classical set theory affects reasoning and analysis procedures, adding all the problems of an "early and precisely classification". The final decision is made after steps which drastically reduce the intermediate results. Any constraint is accompanied by an absolute threshold value and no exception is allowed.

As one of the approaches for finding a methodology for multicriterial land use valorization, application of fuzzy logic and genetic algorithm gave acceptable results. Weights used in the initial equations of mathematical presentation, are indirectly optimized by modified objective function applied in GA during fitness values evaluation. Populations of binary vector strings which indirectly represent solutions for criterial weights are maintained by the unique mechanism of GA. Using the applied methodology it is possible to search the alternatives of spatial organization for given land use categories and finding the optimum alternative.

## References

1. Goldberg, D.E.: Genetic Algorithms. Addison Wesley, Reading (1989)
2. Biggero, L., Laise, D.: Organizational Behavior and Multicriterial Decision Aid. In: 2nd Annual Conference of Innovative Research in Management, Stockholm (2002)
3. Matthews, K.B., Craw, S., Mackenzie, I., Elder, S., Sibbald, A.R.: Applying Genetic Algorithms to Land Use Planning. In: Proceedings of 18th Annual Conference of the BCS Planning and Scheduling SIG, ISSN 1368-5708 (1999)
4. Stewart, T.J., Janssen, R., Van Herwijnen, M.: A genetic algorithm approach to multiobjective land use planning. University of Cape Town, University for Environmental Studies, Vrije Universiteit Amsterdam (2004)
5. Ponjavic, M., Avdagic, Z., Karabegovic, A.: Applying Genetic Algorithm to Land Use Planning Problem of Multicriterial Optimization. In: ICAT 2005 - XX International Symposium on Information Communication and Automation Technologies, Sarajevo (2005)
6. Stefanakis, E., Sellis, T.: A DBMS repository for the application domain of GIS. In: 7th International Symposium on Spatial Data Handling, Delft, The Netherlands, pp. 3B19–3B29 (1996)
7. Sasikala, K.R., Petrou, M., Kittler, J.: Fuzzy classification with a GIS as an aid to decision making, University of Surrey, Guildford, Surrey, UK (1996)
8. Karabegovic, A., Konjic, T., Atic, V.: Implementation of Geographic Information System in Electro distribution Tuzla. In: BH K CIGRÉ - International Council on Large Electric Systems - VI Conference, Neum, BiH (2003)
9. Karabegovic, A., Ponjavic, M., Konjic, T.: Geographic Information Systems - a platform for designing and development of Cable Television. In: IKT 2003 - XIX International Symposium on Information and Communication Technologies, Sarajevo (2003)
10. Karabegovic, A., Ponjavic, M.: Informatics support in designing local loops. In: BIHTEL 2004 - V International Conference On Telecommunications, Sarajevo (2004)
11. Karabegovic, A., Avdagic, Z., Ponjavic, M.: Applications of Fuzzy Logic in Geographic Information Systems for Multiple Criteria Decision Making. In: CORP 2006, 11th International Conference on Urban Planning & Regional Development in the Information Society, Vienna (February 2006)

12. Ponjavic, M., Avdagic, Z., Karabegovic, A.: Geographic Information System and Genetic Algorithm Application for Multicriterial Land Valorization in Spatial Planning. In: CORP 2006, 11th International Conference on Urban Planning & Regional Development in the Information Society, Vienna (February 2006)
13. Haupt, R.L., Haupt, S.E.: Practical Genetic Algorithms. John Wiley & Sons, Inc., Hoboken (2004)
14. Haining, R.: Spatial Data Analysis: Theory and Practice. Cambridge University Press, Cambridge (2003)
15. Avdagić, Z.: Vještačka inteligencija i fuzzy – neuro – genetika, GrafoArt Sarajevo (2003)
16. Zhu, A.X., Hudson, B., Burt, J., Lubich, K., Simonson, D.: Soil Mapping Using GIS, Expert Knowledge, and Fuzzy Logic. Soil Sci. Soc. Am. J. 65, 1463–1472 (2001)
17. Longley, P., Goodchild, M., Maguire, D., Rhind, D.: Geographic Information Systems and Science. John Wiley&Sons, Ltd., England (2002)
18. Ross, T.J.: Fuzzy Logic with Engineering Applications. John Wiley & Sons, Chichester (2004)
19. Petry, F.E., Robinson, V.B., Cobb, M.A.: Fuzzy Modeling with Spatial Information for Geographic Problems. Springer, Heidelberg (2005)

# 12

# Searching Multimedia Databases Using Tree-Structures Graphs

Anastasios Doulamis[1] and George Miaoulis[2]

[1] Technical University of Crete, Chania, University Campus, Greece
Tel.: + 30 28210 37430
adoulam@ergasya.tuc.gr
[2] Technological Education Institute of Athens Department of Informatics Ag.Spyridonos St., 122 10 Egaleo, Greece
Tel.: (+30) 2 10 53 85 312; Fax: (+30) 2 10 59 10 975

**Abstract.** This paper presents a system architecture and the appropriate algorithms for confidential searching of multimedia digital libraries. The proposed scheme uses Middleware service layer that allows *pre*-processing of raw content with technology owned by the Search Engine, without compromising the security of the original architecture in any way. The specific search algorithms described are a hierarchical graph structure algorithm for preprocessing, and a backtracking search algorithm that achieves good real-time performance (speed, and precision-recall values) under the given security constraints.

## 12.1  Introduction

Multimedia digital libraries are emerging at an increasingly fast rate throughout the world. There now exist commercial digital archives containing several tens of millions of hours of films and video recordings. This vast amount of multimedia information requires new methods and tools that allow quick searching, indexing and retrieval of audio-visual information in a secure and trusted environment. Efficient and secure searching in multimedia digital libraries requires a) pattern recognition technologies for *analyzing* and *describing* content under a semantic framework, b) knowledge engineering algorithms for quick *searching* of annotated multimedia content, and c) computer science protocols and architectures for protecting *intellectual property rights* of both content and search technologies. In this paper, we address the latter two of these issues, assuming a known multimedia description scheme *stemming from the MPEG-7 standard*. Other description schemes can be also used.

The traditional way of representing a video, as a sequence of consecutive frames, (each of which corresponds to a constant time interval, e.g., 40 ms for the PAL system), while being adequate to "play" a video file in a movie mode, is not appropriate for new multimedia services, such as searching, retrieving and mining of video content over distributed multimedia platforms.

Sequential searching is a time consuming process since video archives contain enormous amounts of information. Thus, searching algorithms for multimedia content

D. Plemenos, G. Miaoulis (Eds.): Arti. Intel. Techn. for Comp. Graph., SCI 159, pp. 199–214.
springerlink.com                                    © Springer-Verlag Berlin Heidelberg 2009

require nonlinear, normally hierarchical organisation schemes. Hierarchical video organisation is supported by the MPEG-7 standard [1] through the HierarchicalSummary Description scheme [2]. The MPEG-7 standard aims to provide a framework for multimedia description to support efficient searching, retrieval and mining of audio-visual content. The standard provides an XML schema for encoding hierarchical video organisation tools. It also suggests an algorithm for the implementation of a hierarchical video summarisation scheme, by extracting key-frames and then clustering the remaining video frames according to the key-frames' visual content and the temporal distance between the key-frames and the remaining video frames [2].

Specific algorithms for improving the searching performance in large scaled databases have been surveyed in the last years from the knowledge research community. Examples include the B-trees structures or modified versions of them through $B^+$-trees, R-trees and KD-trees [3]. However, B and $B^+$ trees use single valued keys which are not appropriate for indexing multi-dimensional multimedia structured. Despite the fact that some of them, they can apply for multi-dimensional structures, their performance still remains problematic due to the "curse of dimensionality" phenomenon in which data are represented as points in the multi-dimensional feature space undergoing noise [4]. It has been shown in Weber et. al. [5] that when the dimension increases beyond ten, some of these indices yield worsen performance than the sequential search.

The work of [6] uses Order Vector Approximation File Techniques (OVA-FILE) for indexing video data. This method partitions the feature files into slices such that only a small number of slices are accessed and checked during k Nearest Neighbour (kNN) search. The authors [7] introduce a new graph-theoretical clustering algorithm on large databases of signatures. A Hierarchical Cellular Tree (HCT) is presented in to bring an effective solution especially for indexing large multimedia databases [4].

In the VideoZoom [8] prototype, video frames are linearly decomposed in space and frequency domains to allow for fast and precise video browsing. VideoZoom is based on a linear spatiotemporal video decomposition. Content-based video organisation is achieved in [9] by hierarchically analyzing a video sequence into different content resolution levels, using a tree structure representation, the levels of which indicate the respective content resolution, while the nodes correspond to the segments into which the video is partitioned at this level. Based on this scheme, the user is able to select segments (tree-nodes) of interest and reject segments of non-interest, resulting in a multi-resolution *interactive* video browsing schema. In [10], a video abstract is proposed while a pictorial summary of the multimedia content is discussed in [11]. Finally, the cluster validity method is presented in [12] for hierarchical video content summarization.

While the aforementioned approaches are suitable for video browsing applications, they can not be directly applied for video searching, mining and retrieving scenarios since the algorithms are performed interactively, with a user guiding the process. In particular video content is traced by the user's interaction and not in an automatic way, as it is required for the secure searching and mining schemas.

In the context of this paper, we propose a platform, which can support multimedia searching where both preprocessing and real-time searching are performed in an automatic way, i.e., without any user interaction. In particular, we propose a preprocessing scheme which uses a hierarchical graph structure able to organize video data into different content resolution levels, resulting in a pyramidal hierarchy from the coarsest (lowest) to finest (highest) content resolution. Instead of the *tree* structure organization of [9], in the presented work the nodes are associated with each other, resulting in a pyramidal *graph* representation. This extension is the innovation that allows for automatic tracing of multimedia content, enabling offline, non-interactive execution. In contrast, in the tree representation of [9], video content is traced by the user's interactions, making the scheme suitable only for video *browsing* applications.

We propose a new algorithm for real-time multimedia search, based on the preceding preprocessing scheme. The algorithm comprises two different phases: forward search and backtracking. In the forward search, the precomputed hierarchical graph is examined to find best matched nodes with respect to the user's query. In the backtracking phase, different alternative paths of the hierarchical graph are returned in order to increase retrieval precision.

## 12.2   Hierarchical Graph Representation of the Multimedia Content

To enable multimedia queries, we need to describe the rich media content through the use of appropriate multimedia metadata. Multimedia metadata are extracted by the application of audio-visual signal processing algorithms and stored in separate databases in Middleware layer. When a specific query is to be executed, these metadata are exploited to get the retrievals instead of the raw data. Multimedia metadata are offline estimated. In the proposed scheme the extracted multimedia metadata are organized in hierarchical graph structures to allow for the application of real-time search algorithms.

### 12.2.1   Overview of the Proposed Multimedia Search Architecture

The presented architecture used for multimedia content search over distributed and heterogeneous databases consists of two modules. The first is responsible for re-organizing the content into an indexed structure that will allow a quick but efficient search for relevant multimedia content over the raw multimedia databases. This module is named multimedia content organization using tree-structured graphs since its basic concept is to decompose video sequences into different levels of content hierarchy and structure the multi-dimensional multimedia indices in such a way that will allow fast retrieval of relevant information. The second module is responsible for describing a new multi-dimensional search algorithm that exploits the above mentioned structure and yields quick but reliable retrieval performance upon a user's query. This module requires low computational complexity since it is applied in a on-line search mode.

The proposed architecture is shown in Fig 12.1. In this figure, we consider *N* distributed multimedia repositories. In each repository different multimedia data have been stored as raw material. Over these data, the off-line tree-structure graph

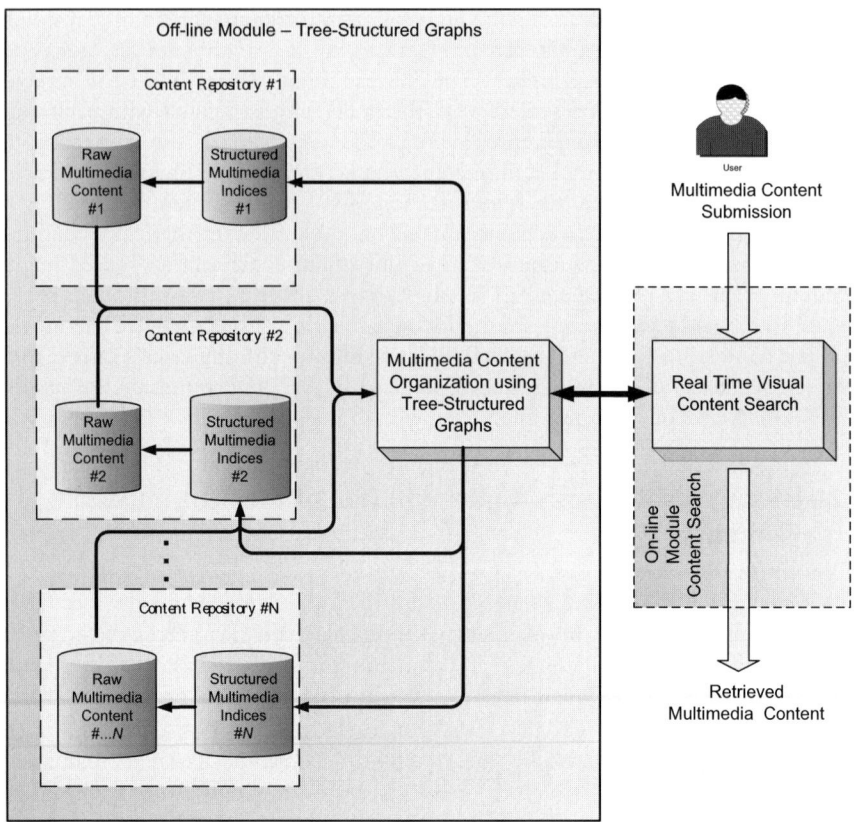

**Fig. 12.1.** A graphical overview of the proposed hierarchical graph structure for multimedia content representation and quick but efficient search

construction algorithm is initially activated to generate structured multimedia indices, which will allow a quick but efficient search of the multimedia database. Upon these structured indices, the real time search algorithm is applied. The algorithm exploits the structured indices and searches for relevant multimedia content by navigating through the structured graphs.

### 12.2.2  Hierarchical Multimedia Content Representation

In this section, we describe the algorithm that is used during the offline operation phase to generate the multimedia metadata. The proposed scheme used to non-linearly organize the multimedia content in a way that supports quick search. This hierarchy is encoded using the XML schema of the MPEG-7 standard to guarantee interoperability and universal accessibility. Note that this algorithm is executed entirely noninteractively instead of the previous approaches such as the work of [9].

**Fig. 12.2.** The proposed hierarchical graph structure for multimedia content representation

In this paper, a *hierarchically structured graph* is adopted to nonlinearly organise the content of a video file. This is an important enhancement of the tree structure adopted in [9] since it allows automatic search without user's interaction. The depth of the graph indicates different content resolution levels starting from the coarsest (lowest) and ending with the finest (highest) resolution. Three different resolution levels are adopted in the presented approach; the *shot representative level,* the *shot level* and the *frame representative level.* At the *shot representative level,* the content of a video file is projected on the key-shots space. This means that at this level, the nodes of the graph correspond to the shot representatives of a video file. Similar content organisation is accomplished at the third resolution level (*frame representative level*) with the difference that video content is projected on the key-frames instead of the key-shots. Therefore, the graph nodes of the third resolution level are the frame representatives. Finally, at the second resolution level (*shot level*), video content is represented by the shot information. Shots are nonlinearly organised with respect to the shot representatives. At each resolution level, links are assigned from one node to another, which indicate the *degree of association* between the connected nodes. In this way, a node is related with other nodes of similar content characteristics but at the same content resolution level. Fig. 12.2. presents an example of the proposed three-level hierarchically structured graph adopted for representing video content in a nonlinear hierarchical way.

Shot and frame representatives are extracted by minimising a cross correlation criterion so that the ones that correspond to the most "uncorrelated" content are defined as shot/frame representatives. Instead, the *shot level* is constructed by applying a clustering algorithm.

### 12.2.3   Shot/Frame Representative Level Construction

In this section, we describe the way of extracting shot and frame representatives to construct the graph nodes of the first and third resolution levels.

Let us denote as $\mathbf{f}_i$ the vector, the elements of which correspond to the features exacted either for the $i^{th}$ frame or the $i^{th}$ shot of a video sequence. Let us also denote as $K$ the number of frame or shot representatives being adequate to describe the rich content fluctuation. The number $K$ can be estimated as in [9] so that the *"difficulty"*, measured as the number of frames that must be processed before finding content that fulfills the user requirements, is minimised.

Let us now denote as $\mathbf{x}=[x_1,x_2,\ldots,x_K]^T$ a vector, the elements of which refer to the indices of the $K$ representatives. These indices are found in this paper by minimising a cross correlation criterion, as

$$\hat{\mathbf{x}} = \arg \min_{\mathbf{x}} E(\mathbf{x})$$

$$E(\mathbf{x}) = \frac{2}{K(K-1)} \sum_{i=1}^{K-1} \sum_{j=i+1}^{K} \rho(\mathbf{f}_{x_i}, \mathbf{f}_{x_j})^2 \qquad (12.1)$$

where $\rho(\mathbf{f}_{x_i}, \mathbf{f}_{x_j})$ is the correlation coefficient of feature vectors $\mathbf{f}_{x_i}$ and $\mathbf{f}_{x_j}$ at the indices $x_i$ and $x_j$ respectively, while $\hat{\mathbf{x}}$ gives the optimal index vector. Equation (12.1) indicates that the representatives should be as much as possible uncorrelated with each other.

The complexity of an exhaustive search for estimating the minimum value of (12.1) would be unreasonably great, since all possible combinations of frames would need to be examined. For this reason, minimisation of (12.1) can be applied by the use of a genetic algorithm as in [9]. This scheme is able to find a solution close to the optimal one within a small number of iterations.

### 12.2.4   Shot Level Construction

Having estimated the most representative shots and frames within a video sequence, the following step of the proposed approach is to construct the shot classes so that the second level of the proposed video hierarchy is developed.

Let us now assume that the $K$ most uncorrelated shots have been estimated using the aforementioned algorithm, that is the optimal index vector $\hat{\mathbf{x}}$. Then, the $S_k$ with $k=1, 2, ..,K$ shot classes are constructed as

$$S_k = \{s_i : i \in Z(\hat{x}_k)\} \text{, for all } i \tag{12.2}$$

where $Z(x_k)$ refers to the influence zone of the index $x_k$. (i.e., the $k^{\text{th}}$ shot representative) and $s_i$ to a video shot. The influence zone is defined as a set which contains all shot indices, whose respective shot feature vector is closer to the feature vector of the representative shot defined by the index $x_k$ than all the other representative shots, that is

$$Z(x_k) = \{\forall i : \rho(\mathbf{f}_i, \mathbf{f}_{x_k}) > \rho(\mathbf{f}_i, \mathbf{f}_{x_m})$$
$$\forall m \in \{1, 2, \cdots, K\} \text{ and } m \neq k\} \tag{12.3}$$

Equation (12.3) means that all shots that fall within the same influence zone of a shot representative are considered as members of the same class.

### 12.2.5  Node Association

In our scheme, instead of the approach of [9], each node of the graph is associated with the other nodes of the same level with similar visual content characteristics. Such an association yields a hierarchically structured graph representation scheme which in the sequel significantly increases the mining and searching efficiency. Modification of the algorithm presented in [9] to support association among nodes of the same content resolution level is performed to allow for automatic searching. Instead, the method of [9] is oriented only for interactive video navigation that demands user feedback. The necessity of node association is described in Section 12.3, in which we present the searching algorithm.

Node association is performed using the correlation coefficient $\rho(\cdot)$ between nodes that are to be linked. More specifically, let us denote as $n_i(k)$, and $n_j(k)$ two nodes of the hierarchical graph at the $k^{\text{th}}$ content resolution level. Then, between these two nodes a link is assigned with similarity degree the cross correlation of the feature vectors related with the content of these nodes, i.e.,

$$d(n_i(k), n_j(k)) = \rho(\mathbf{f}_{n_i(k)}, \mathbf{f}_{n_j(k)}) \tag{12.4}$$

where $d(n_{i(k)}, n_{j(k)})$ expresses the similarity degree among the nodes $n_i(k)$, and $n_j(k)$, while $\mathbf{f}_{n_i(k)}, \mathbf{f}_{n_j(k)}$ indicate the respective feature vector of the nodes $n_i(k)$, and $n_j(k)$.

The similarity degrees of a node are sorted in descending order so that the first links point out for nodes of the most similar content with respect to the reference node. As a result, in case that searching succeeds at some nodes at a given level, the searching can then follow nodes of similar content to get more relevant results to the user.

In particular, let us denote as $Assoc(n_i(k))$ a set which contains the sorted associate nodes of the node $n_i(k)$. Thus, the $j^{th}$ associate $a_j(n_i(k)) \in Assoc(n_i(k))$ refers to the $j^{th}$ most relevant node with respect to $n_i(k)$. Consequently, $a_0(n_i(k))$ is the best associate node of $n_i(k)$ based on the similarity degree described in equation (12.4).

## 12.3 Real Time Search

In this section, we describe a non-sequential content navigation algorithm that can be used for the online, real-time search phase of the system. The proposed algorithm exploits the hierarchically structured graph described in Section 12.2, which nonlinearly organises video content. The process is fully automatic and is characterised by small computational complexity compared to sequential (linear) searching, while simultaneously achieving high precision-recall performance.

The proposed automatic searching algorithm is divided into two main parts: *forward searching* and *backtracking*.

### 12.3.1 Forward Searching

The purpose of the forward searching is to seek all nodes beneath a node given as input in the process and navigate towards relevant video content. The input node of the forward searching process is given from the backtracking process as described in Section 12.3.4 (but see Section 12.3.5 for initialization and termination of the algorithm). In our approach, we assume that the relevant content is located only at the third resolution level, that is at the *key-frame representatives*, so that only references to the most detailed content level are returned to the user as a final result (the other levels do serve for navigating the content efficiently, of course).

The forward searching process is divided into two main procedures: *best node selection* and *node retrieval*.

### 12.3.2 Best Node Selection

The *best node selection* procedure takes as input a node $n$ and the respective $k$ content resolution level at which node $n$ is located. The purpose of this procedure is to seek all nodes beneath node $n$ in order to *return* the node that best matches the user's query and has not been previously selected. In particular, let us denote as $Child(n)$ the set which contains all the children of $n$. Variable $c_i \in Child(n)$ expresses the $i^{th}$ child of $n$. Then, the best node at the following $k+1$ resolution is the child of $n$ that best matches to the user's query and has not been previously returned.

$$\hat{c} = \arg\max_{c_i \in Child(n),\, c_i \notin R} \{\rho(\mathbf{f}_q, \mathbf{f}_{c_i})\} \qquad (12.5)$$

where $R$ is a set which contains all the processed nodes for a given user's query. Variable $\hat{c}$ refers to the best matched child, while $\mathbf{f}_q$ and $\mathbf{f}_{c_i}$ to the query feature vector and the feature vector of the child $c_i$. We recall that $\rho(\cdot)$ is the correlation coefficient among the two feature vectors.

Having selected the best matched child, the process iteratively proceeds until a) either the last content resolution level is reached (i.e., $k= 3$) or b) all children have been previously returned or there are no children for the examined node. At the termination of the procedure the best matched node, say $\hat{n}$ is returned.

The following table summarizes the main step of the *best node selection at the next iteration*.

### 12.3.3  Node Retrieval

This procedure is activated in case that $k=3$, i.e., the best node selection procedure reaches *the key-frame representative level*. The procedure takes as input the best matched node $\hat{n}$, as provided from the *best node selection* procedure and returns as output the modified set $R$ by adding in it a set of nodes that are considered relevant to the user's query.

More specifically, the best matched node $\hat{n}$ is included in the set $R$. Apart from this node, set $R$ is enhanced with those associate nodes of $\hat{n}$ which satisfy the following equation

$$a_r(\hat{n}) \in Assoc(\hat{n}):$$
$$\gamma_r \cdot \rho(\mathbf{f}_q, \mathbf{f}_{a_r(\hat{n})}) > \rho(\mathbf{f}_q, \mathbf{f}_{\varphi(\hat{n})}) \tag{12.6}$$
$$\text{and } a_r(\hat{n}) \notin R$$

where $a_r(\hat{n})$ is the $r^{\text{th}}$ best sorted associate of the node $\hat{n}$ and $\varphi(\hat{n})$ the respective father node. Vectors $\mathbf{f}_q, \mathbf{f}_{a_r(\hat{n})}$ and $\mathbf{f}_{\varphi(\hat{n})}$ indicate the feature vectors of the query, $a_r(\hat{n})$ node and the father node of $\hat{n}$, i.e., the $\varphi(\hat{n})$. Variable $\gamma_r$ scales the correlation according to the order of the associate nodes. In our approach, $\gamma_r$ increases exponentially with respect to the order $r$

$$\gamma_r = \beta^r, \ \beta > 1 \tag{12.7}$$

Equation (12.7) means that all the associate nodes $a_r(\hat{n})$ of $\hat{n}$, whose correlation coefficient between the query and the node $a_r(\hat{n})$ (scaled by a factor $\gamma_r$ exponentially proportional to the rank $r$ of the associate node) is greater than the respective coefficient between the query and the father node $\varphi(\hat{n})$ of $\hat{n}$, are included in the set $R$, containing the relevant nodes of the query. This is performed only for those nodes that have not been previously selected as relevant in set $R$.

### 12.3.4  Backtracking

After the completion of the forward searching process, backtracking is activated to examine other possible paths (see Section 12.3.5 for initialization and termination of the algorithm). The backtracking procedure takes as input the father of the node $\hat{n}$, i.e., the $\varphi(\hat{n})$, as obtained from the termination of the *BestNodeSelection* procedure. The algorithm initially finds the first associate node of $\varphi(\hat{n})$ that has not been previously selected (i.e., it does not belong to the set $R$), that is

$$\alpha \equiv a_{r_o}(\varphi(\hat{n})) \tag{12.8}$$

$$\text{with } r_0 = \arg\min_r a_r(\varphi(\hat{n})) \notin R \tag{12.9}$$

Let us denote as $D$ the difference of the correlation coefficient between the query and node $\alpha$ from the coefficient between the query and the father of $\alpha$, i.e., $\varphi(\alpha)$,

$$D = \rho(\mathbf{f}_q, \mathbf{f}_\alpha) - \rho(\mathbf{f}_q, \mathbf{f}_{\varphi(\alpha)}) \tag{12.10}$$

In case that $D \geq 0$ the backtracking process terminates and the forward searching is activated with input the node $\alpha$ along with the respective content resolution level. On the contrary, in case of $D < 0$, indicating that the father of node $\alpha$ is more relevant to the user's query than node $\alpha$, another backtracking is activating with input the $\varphi(\alpha)$. The process is terminated whether either $D \geq 0$ or the root of the graph is reached.

Equation (12.10) can not be calculated in case that node $\alpha$ is a root node, since the father $\varphi(\alpha)$ does not exist. This case means that another graph is examined, i.e., another video file. This file is the first associate of the current examined video file. In this scenario, the forward searching is directly activated with input the root of the new graph, that is the $\alpha$.

### 12.3.5  Initialization and Termination

The search algorithm starts by activating the Best Node Selection procedure of the forward searching. In particular, all the graph roots of video files belonging to the same category of the query are examined and the one that best matches the query is initially selected for further searching. Let us denote as $n_i(0)$, $i=1,2,\ldots$ the root nodes of graphs for video files belonging to the same category as the query one. Then, the best matched root node is found as the one which maximizes the

$$n_{\hat{k}}(0) = \arg\max_{n_i(0)} \rho(\mathbf{f}_q, \mathbf{f}_{n_i(0)}) \tag{12.11}$$

The node $n_{\hat{k}}(0)$ along with the respective zero content resolution level are used as input in the *BestNodeSelection*( $n_{\hat{k}}(0)$,0) for commencing the search algorithm.

The search algorithm terminates in case that the number of elements of the processed frames in $R$ reaches a maximum number of data, that is $|R| = N$, where $N$ is the number of processed frames. The best $K$ of these ( $K \leq N$ ), either based on the correlation coefficient itself or on the results of the per-frame processing, are delivered as actual search hits.

## 12.4   Experimental Results

In order to evaluate the efficiency of the proposed hierarchical tree-structure video representation for multimedia content searching, and to compare it with other approaches presented in the literature, objective quality criteria need to be introduced. A series of objective criteria are adopted in this paper for evaluating search performance. The criteria are the Precision-Recall curve, the Average Normalised Modified Retrieval Rank (ANMRR), and the Search Efficiency Ratio [13][14] [15].

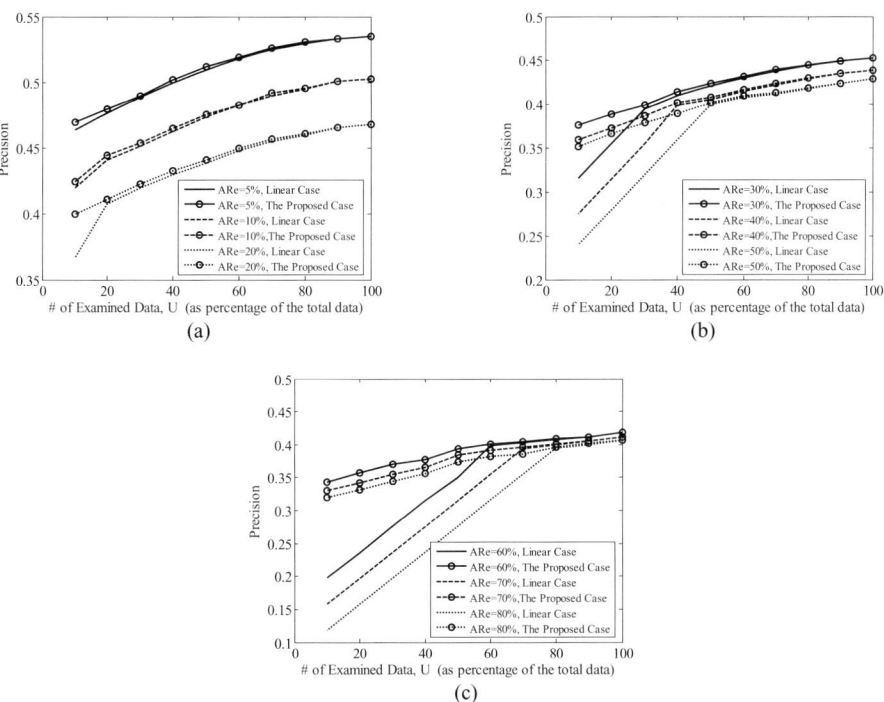

**Fig. 12.3.** Comparison of precision values of the proposed scheme with the linear (sequential) case, for varying proportions of examined data, expressed as percentage of the total data $U$. (a) Recall values of 5%, 10%, and 20%. (b) Recall values of 30%, 40%, and 50%. (c) Recall values of 60%, 70%, and 80%.

To evaluate the efficiency of the proposed scheme, we have used multimedia database consisting of 30 video files, each of average duration of 2 hours. The experiment is conducted by submitting 3,000 randomly selected queries and then computing the aforementioned described the objective criteria. The proposed scheme is initially compared with the traditional linear (sequential) search, which is currently the most popular multimedia search algorithm. We also present comparisons with other nonlinear organisation schemes, such as the works of [8] [9][10][11] and [12]. It should be clarified, however, that the approach of [9] assumes a user's interaction in decomposing the visual data and thus it can not be directly applied in the case where automatic video organisation is required. For this reason, we compare the proposed approach with these methods under the SER ratio, which does not require precision accuracy values.

Fig. 12.3 presents the precision values versus the number of examined frames (data) $U$ in the database. In this case, we have omitted the subscripts $s$ and $n$, since the experiments have been conducted for the same value of $U$ both for the sequential and the proposed search approach. In addition, in Fig. 12.3 variable $U$ is expressed as percentage of the total number of data in the database for convenience. The experiment has been conducted for different recall values and for factor $\beta =1.1$ [see equation (12.7)].

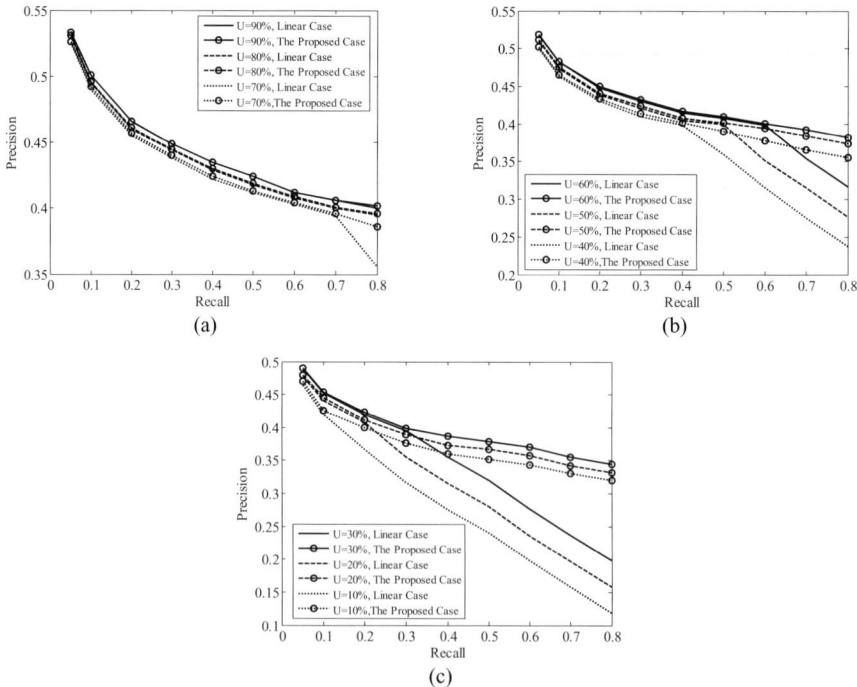

**Fig. 12.4.** Comparison of the precision-recall values of the proposed scheme and of the linear (sequential) case. (a)$U$=90%, 80%, and 70%. (b) $U$=60%, 50%, and 40%. (c) $U$=30%, 20%, and 10%.

These results show that, in terms of precision, our system very slightly outperforms the sequential case as long as the number of examined frames $U$ is greater than the recall target. For values of $U$ lower than the targeted recall, an almost linear fall of the precision values for the linear approach occurs whereas our system suffers only small precision degradation. This is explained as follows. Let us assume that the content is uniformly distributed across all data of the database (an assumption that perfectly fits the random location of the data). Then, we can see that we are able to find an adequate number of relevant items, as long as $U$ is greater than the recall value, both for the proposed and for the linear case. This means that we examine a sufficient quantity of data in the database to reach the recall target. On the other hand, in case that the number of examined data becomes less than the recall target, we can see that the examined number of samples is not on average adequate to reach the recall value for the linear case leading to a significant deterioration of the precision performance. In contrast, the performance of the proposed method remains very robust since the algorithm navigates the database in regions that present high probability of locating relevant data. Thus, it can successfully discover content even if it can only examine a small proportion of the total data available. The rapid decrease for the linear case is more evident for high recall values. Note that the best performance achievable is in any case limited from above by the intrinsic level of difficulty at the individual frame level of extracting discriminating features from the content of interest; this means that the exact values shown in these figures are application-specific and thus, as far as. evaluating the performance of our system is concerned, arbitrary – it is the comparison between the results of our method and those of the linear case that is important.

The conclusions drawn above are also supported by the results shown in Fig. 12.4, where we have plotted the precision-recall curve for different values of $U$ both for the

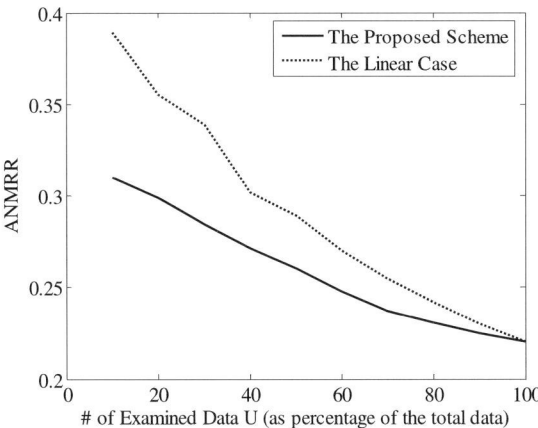

**Fig. 12.5.** Performance comparison of the ANMRR values of the proposed scheme with the linear (sequential) case versus the proportion of examined data, expressed as percentage of the total data $U$

proposed method and for the linear approach. It was again clear that, in the linear case, the precision accuracy drops abruptly when $U$ exceeds the recall values. On the other hand, the precision accuracy in the proposed scheme remains robust regardless of the ARe and $U$ values. Once again, factor $\beta = 1.1$ is selected.

The efficiency of the proposed scheme as far as the ANMRR values are concerned is depicted in Fig. 12.5 versus the number of examined data $U$. This figure compares the results obtained from the proposed and the sequential case. The figure demonstrates that in the proposed scheme the ANMRR values remain relatively stable regardless of $U$. On the contrary, in the linear (sequential) case the ANMRR values drop sufficiently only as the proportion of examined data U increases considerably.

The results obtained using the Search Efficiency Ratio (*SER*), as defined in equation (12.20) for different non-linear approaches are presented in Fig. 12.6. The *SER* criterion can be used to compare the performance of other non-linear search methods presented in the literature than the proposed one. In particular, Fig. 12.6(a) compares

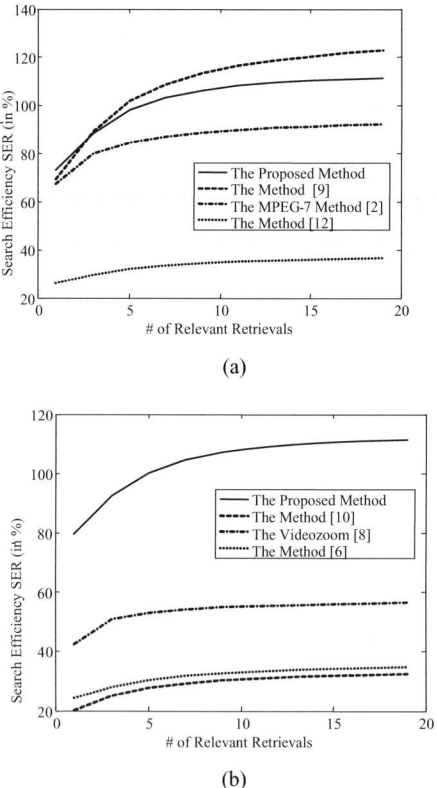

(a)

(b)

**Fig. 12.6.** The Search efficiency ratio (SER) versus the number of Relevant Retrievals. (a) A comparison with the methods of [2],[9],[12]. (b) A comparison with the methods of [8],[10],[11].

**Table 12.1.** The SER ratio of the proposed scheme compared with other approaches

| Nonlinear Video Representation Algorithms | SER (1st Retrieval) | SER (5th Retrievals) | SER (15th Retrievals) |
|---|---|---|---|
| The Proposed Scheme | 79.60 | 100.32 | 110.56 |
| The Method of [9] | 87.20 | 108.84 | 121.35 |
| The Method of [2] (MPEG-7) | 67.42 | 84.63 | 91.36 |
| The Method of [10] | 20.18 | 27.80 | 31.92 |
| The Method of [8] | 42.30 | 53.07 | 56.02 |
| The Method of [11] | 24.20 | 30.50 | 34.33 |
| The Method of [12] | 26.13 | 32.19 | 36.14 |

the performance of the proposed approach versus the number of retrievals being considered as relevant for the methods of [2],[9],[12], while Fig. 12.6(b) for the methods of [8],[10],[11]. We can conclude that the presented search methods reaches the performance of the [9] though the latter is an interactive method that a user guides the process, while the proposed one is an automatic approach. For all the other compared approaches, interactive or not, the presented one yields better performance, revealing its efficiency.

Finally, the results obtained at given values of the number of retrievals using the *SER* criterion for all the above mentioned compared approaches are shown in Table 12.1. As is observed, the proposed video hierarchy provides a significant reduction of the *difficulty* in accessing frames of interest compared to sequential scanning (about 79 times for the first relevant retrieval). In this table, we have also compared the performance of the proposed algorithm with other hierarchical approaches for video content decomposition and navigation, despite the fact that these approaches can not be used for multimedia data mining.

In Table 12.1 we also compare the proposed scheme to the semi-automatic method of [9],and the performance of the latter is in fact better. This is due to the fact that the work of [9] is semi-automatic, requiring user feedback for decomposing video content.

## 12.5  Conclusions

In the proposed architecture, a non-linear organization of multimedia data is presented that allows the direct implementation of an automatic non-linear search module. The algorithm, instead of the previous work of [9], which uses a tree-based decomposition of the data, exploits the application of a pyramidal graph that allows search to be applied directly on the content domain.

An automatic non-linear search algorithm is applied in the following to exploit the advantages of the proposed graph-structure hierarchical media content organization.

Experimental results using a series of objective criteria as well as comparisons with other approaches have been proposed to demonstrate the efficiency of the presented scheme than previously published works. The results have been obtained on a very large media database.

# References

[1] MPEG-7 Requirements Group, MPEG-7: Context, Objectives and Technical Roadmap, vol.12, Vancouver, ISO/IEC SC29/WG11 N2861 (July 1999)

[2] ISO/IEC, J.T.C.: 1/SC 29/WG 11/N3964, N3966, Multimedia Description Schemes (MDS) Group, Singapore (March 2001)

[3] Lu, G.: Techniques and Data Structures for Efficient Multimedia Retrieval Based on Similarity. IEEE Trans. Multimedia 4(3), 372–384 (2002)

[4] Kiranyaz, S., Gabbouj, M.: Hierarchical Cellular Tree: An Efficient Indexing Scheme for Content-Based Retrieval on Multimedia Databases. IEEE Trans. Multimedia 9(1), 102–119 (2007)

[5] Weber, R., Schek, H.-J., Blott, S.: A Quantitative Analysis and Performance Study for Similarity-search Methods in High-dimensional Spaces. In: Proc. of the 24rd Int. Conf. Very Large Databases, pp. 194–205, August 24-27 (1998)

[6] Lu, H., Chin Ooi, B., Tao Shen, H., Xue, X.: Hierarchical Indexing Structure for Efficient Similarity Search in Video Retrieval. IEEE Trans. on Knowledge And Data Engineering 18(11), 1544–1559 (2006)

[7] Cheung, S.-c.S., Zakhor, A.: Fast Similarity Search and Clustering of Video Sequences on the World-Wide-Web. IEEE Trans. on Multimedia 7(3), 524–537 (2005)

[8] Smith, J.R.: VideoZoom: Spatio-temporal video browser. IEEE Trans. on Multimedia 1(2), 157–171 (1999)

[9] Doulamis, A., Doulamis, N.: Optimal Content-based Video Decomposition for Interactive Video Navigation over IP-based Networks. IEEE Trans. on Circuits and Systems for Video Technology (to appear June, 2004)

[10] Nam, J., Tewfik, A.H.: Video Abstract of Video. In: Proc. of the IEEE Inter. Workshop on Multimedia Signal Processing, pp. 117–122, Copenhagen, Denmark (September 2000)

[11] Yeung, M.M., Yeo, B.-L.: Video visualization for compact presentation and fast browsing of pictorial content. IEEE Trans. on Circuits and Systems for Video Technology 7(5), 771–785 (1997)

[12] Hanjalic, A., Zhang, H.: An integrated scheme for automated abstraction based on unsupervised cluster-validity analysis. IEEE Trans. on Circuits and Systems for Video Technology 9(8), 1280–1289 (1999)

[13] Salton, G., McGill, M.J.: Introduction to Modern Information Retrieval. McGraw-Hill Book Company, New York (1982)

[14] MPEG-7 Visual part of eXperimentation Model Version 2.0, MPEG-7 Output Document ISO/MPEG (December 1999)

[15] Doulamis, N., Doulamis, A.: Evaluation of Relevance Feedback Schemes in Content-based in Retrieval Systems. Signal Processing: Image Communications (April 2006)

# Author Index

Printing: Krips bv, Meppel, The Netherlands
Binding: Stürtz, Würzburg, Germany